GINGERS
of PENINSULAR MALAYSIA
AND SINGAPORE

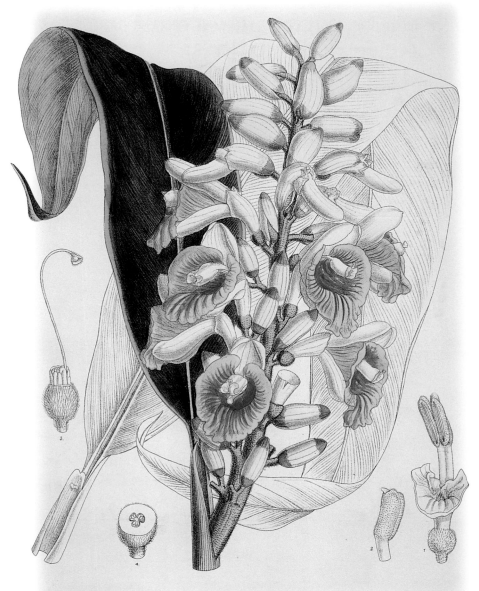

GINGERS
of PENINSULAR MALAYSIA AND SINGAPORE

K. Larsen, H. Ibrahim, S.H. Khaw and L.G. Saw

Edited by
K.M. Wong

with photographs by
Ali Ibrahim, C.L. Chan, J.B. Comber, H. Ibrahim,
S.H. Khaw, A. Lamb, K. Larsen, Alan Ng, L.G. Saw,
E. Soepadmo, I. Theilade, A. Weber and Yong Hoi Sen

Natural History Publications (Borneo)
Kota Kinabalu

1999

Published by

Natural History Publications (Borneo) Sdn. Bhd.
A913, 9th Floor, Wisma Merdeka
P.O. Box 13908
88846 Kota Kinabalu, Sabah, Malaysia
Tel: 6088-233098 Fax: 6088-240768
e-mail: chewlun@tm.net.my

Text copyright © 1999 Natural History Publications (Borneo) Sdn. Bhd.
Photographs and illustrations copyright © 1999 as credited.

All rights reserved. No part of this publication may be reproduced, stored in a retrieval system, or transmitted in any form or by any means, electronic, mechanical, photo-copying, recording, or otherwise, without the prior permission of the copyright owners.

First published 1999

Gingers of Peninsular Malaysia and Singapore
by K. Larsen, H. Ibrahim, S.H. Khaw and L.G. Saw
edited *by* K.M. Wong

Design and layout by C.L. Chan

Frontispiece:	*Alpinia latilabris* (Reproduced with permission of the Royal Botanic Gardens, Kew)
Facing Message:	*Alpinia rafflesiana* (Photo: Yong Hoi Sen)
Facing Foreword:	*Zingiber spectabile* (Photo: Anthony Lamb)
Facing Preface:	*Etlingera elatior* (Photo: C.L. Chan)

Perpustakaan Negara Malaysia   Cataloguing-in-Publication Data

Gingers of Peninsular Malaysia and Singapore / K. Larsen ...
   [et al.] ; edited by K.M. Wong ; photographs by Ali
   Ibrahim ... [et al.].
   Bibliography: p. 127
   Includes index
   ISBN 983-812-025-1
   1. Ginger. 2. Ginger industry. 3. Zingiberaceae. I. Larsen,
   K. II. Wong, K.M. III. Ali Ibrahim.
   584.39

Printed in Malaysia

Contents

Message   vii
Foreword   ix
Preface   xi

Chapter

1: *Introduction*   1
2: *Relationships of the Ginger family*   3
3: *Use and Commercial Importance*   7
4: *Plant Structure in the Ginger family*   12
5: *Pollination and Seed Dispersal*   19
6: *A Brief History of Research on Malayan Gingers*   21
7: *The Tribes and Genera of the Zingiberaceae in Peninsular Malaysia and Singapore*   24
8: *Plant Chemistry in relation to Classification*   93
9: *The Study and Collection of Gingers*   96
10: *Notes from a Ginger naturalist*   99
11: *A Ginger Garden*   110
12: *Peninsular Malaysian and Singapore Gingers —A Checklist*   115

Glossary   122
References and General Reading   127
Index to Scientific and Vernacular Names   132

Message

Natural history publishing is an important, even integral, part of the scientific documentation of any interesting aspect of our living world. In Malaysia, this is on the rise and is a healthy trend.

With so much in the flora and fauna to be studied, documented and then applied for conservation and utilisation, there is an understandable need to emphasize priorities in gaining and presenting this knowledge as it is made available. One critical aspect, surely, is to have the key components of our natural heritage better appreciated by people in general. For the plant and animal life, books on a number of these have now been produced, on the rain forests, orchids, palms, bamboos, proboscis monkeys, orang utan, and so on. Joining this range of titles is the present book on gingers, which are surely (and at the same time) a group of plants that are very important in our lives and yet still little studied in terms of their natural richness in form and species.

I hope this trend continues and that this book on Peninsular Malaysian gingers will interest even more people about the fascination of plants. It should, at least, remind us of how interesting nature is, in its richness and in the way we must depend on it.

Tham Nyip Shen
Deputy Chief Minister,
Sabah

Foreword

Gingers are one of the most amazing components of the plant wealth of Peninsular Malaysia, Singapore and neighbouring countries. Yet there has been a lack of published material easily accessible to non-specialists, which outlines their botany and fascinating development of form, and which also provides an overview of their uses, chemistry and study. Perhaps this is not surprising in a region filled with botanical treasures, for which there is also a general scarcity of taxonomic botanists.

Some of the earlier work documenting the Gingers as a highly diverse plant group in the Malay Peninsula was provided by the great botanists Ridley and Holttum, who carried out much of their research while based in the Singapore Botanic Gardens. This interest has now spread, and it is very gratifying to note that specialists from botanical institutions in Denmark and Malaysia are continuing these studies, making use of the resources—both on record and in living form—brought together at the Singapore Botanic Gardens and the Rimba Ilmu, the botanical garden of the University of Malaya.

Peninsular Malaysia and Singapore have had better basic documentation of their plant biodiversity than many other tropical regions of the world. Yet these areas represent only a part of the overall Southeast Asian tropics, for which much intensive work still remains to be done.

The Gingers have clear ornamental and commercial value, and this book will surely stimulate interest in the region and elsewhere by promoting the study and conservation of these interesting plants.

Tan Jiew Hoe
Singapore

Preface

If biodiversity is, as some put it, the spice of life, then Gingers enhance that very taste. In our rain forests, gingers are an unmistakable life form and their diversity and biology have still not been completely understood or documented. It is a welcome development that scientists do not just record their knowledge in a technical language for specialist journals but also join in to produce overview accounts that bring across the interesting facets of plant diversity to a general readership, and it is with this in mind that this book was produced.

The team comprising the four authors—Professor Kai Larsen of the University of Aarhus, Dr Halijah Ibrahim of the University of Malaya's Institute of Biological Sciences, Ms S.H. Khaw, a botanist studying Malayan gingers, and Dr L.G. Saw (Forest Research Institute Malaysia)—and the editor, Dr K.M. Wong (Institute of Biological Sciences) have collaborated well with Natural History Publications (who have in Mr C.L. Chan, its Managing Director, an equally ardent botanist) to provide the main ingredients for this success. Everyone will also agree that the inclusion of the very fine photographs, especially of Professor H.S. Yong (Institute of Biological Sciences, University of Malaya), Dr L.G. Saw and Dr E. Soepadmo (Forest Research Institute Malaysia), was of the utmost importance in bringing the account "to life".

In the end, one is reminded that a scientific endeavour must have a suitable climate for success. That includes dedicated scientific work, adequate financial commitment and support, as well as an ability to present this work as practical user-oriented packages. Only then can the spice of life be generally appreciated.

Professor Haji Mohamed bin Abdul Majid
Head, Institute of Biological Sciences, University of Malaya
Kuala Lumpur

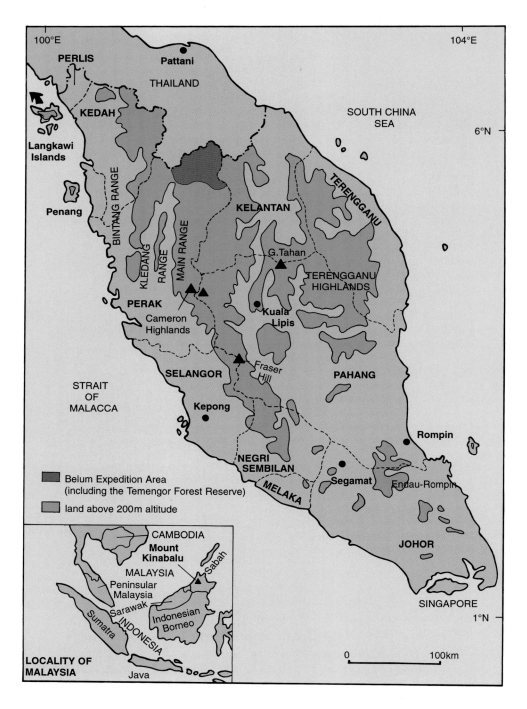

Map of Peninsular Malaysia and Singapore.

A tubular white corolla on a flowering head of *Costus speciosus*.

Extremely common in Peninsular Malaysia and Singapore, *Costus speciosus* displays the spirally arranged leaves and contorted stems that make the Costaceae rather distinct from the Zingiberaceae.

3
Use and Commercial Importance

The main gingers of use come from the genera *Alpinia*, *Amomum*, *Curcuma* and *Zingiber*, and, to a lesser extent, *Boesenbergia*, *Kaempferia*, *Elettaria*, *Elettariopsis*, *Etlingera* and *Hedychium*. At least 20 or more ginger species have been cultivated for their use as spices, condiments, flavours, fresh vegetables (locally known as *ulam*), medicine, ornamentals and quite recently as cut flowers. One of the earliest uses was as spices. The presence of essential oils such as Limonene, Eugenol, Pinene, Geraniol, etc., in many Zingiberaceae species have made some species important since the time of the ancient Greeks.

Rhizomes of the true ginger, *Zingiber officinale*, cultivated throughout the tropics.

There are three species of major commercial importance: *Zingiber officinale* Rosc. (*halia* in Malay), of which the fleshy branched rhizomes are exported to temperate regions from several tropical countries; the turmeric, the rhizomes of *Curcuma domestica* used for colouring food and for medical purposes; and finally, *Elettaria cardamomum* (L.) Maton, the dried capsules of which are exported as cardamom (*buah pelaga* in Malay).

Although grown locally in Malacca since 1416, *Z. officinale* is not native. Its origin is unknown, being cultivated in India and the southern provinces of China from the earliest times. It is, nevertheless, one of the best and the oldest known spices of the Zingiberaceae, and was the basis of lucrative trade during the early times, transported from the East to the Western World. Until today, ginger is still in demand as one of the ingredients in food, bakeries, confectionaries, beverages and traditional medicine. The old rhizomes of *Z. officinale* is used fresh as a flavouring while its young rhizome is eaten raw or pickled, as a relish. It is also sold pickled, candied and dyed red as a sweetmeat in local groceries. This versatile domestic ginger is also used for making ginger beer and gingerbread or biscuit.

Turmeric, known as *Curcuma domestica* in Peninsular Malaysia and *Curcuma longa* in India and some other Asian countries, is popular (after ginger) as a spice used in curries. It is also used as a food flavouring and in ancient times was even exploited as a dye. In comparison to other gingers, turmeric has a long list of uses ranging from spice, flavour, traditional medicine and in cultural beliefs and rites. *C. domestica* is widely grown in the Peninsula and is a native of south-eastern Asia. The broad aromatic leaves of *C. domestica*, sold in bundles in the local market, are used for wrapping fish before steaming or baking. As demand for turmeric is high, it is partially imported along with cardamom.

Cardamom is obtained from the dried fruits and seeds of *Elettaria cardamomum*, indigenous to southern India and Sri Lanka. Its dried fruits are mainly produced in India. It does not grow well in the lowlands of Peninsular Malaysia as it is basically a mountain plant. *E. cardamomum* has been introduced in the Cameron Highlands, but it has not been a commercial success. In some villages in Peninsular Malaysia, it is sometimes possible to find several inferior substitutes of cardamom being planted. Such plants have been identified as *Amomum compactum* or *Amomum kepulaga*.

Species such as *Alpinia conchigera*, *Alpinia galanga* and *Boesenbergia rotunda* are consumed mainly by the local people as spice or condiment. Of the three species, apparently the use of *Alpinia conchigera* and *Boesenbergia rotunda* as spice is restricted to certain areas in the Peninsula. These are frequently cultivated in villages from where they often spread to disturbed habitats, at the fringes of secondary forests or waste land. *Alpinia conchigera*, for instance, is sometimes found quite abundantly along bunds of rice-fields, near ditches and in rubber estates.

Alpinia galanga, generally termed as the greater galangal, although widely cultivated in Southeast Asia, is a minor spice in some Western countries. In Malaysia, its rhizomes are called *lengkuas* and much used in the spicy meat dish called *rendang*. Alternatively, the minor galangal or *Alpinia officinarum*, distributed mainly in China, is used. In Peninsular Malaysia even leaves of some gingers such as those of turmeric, *Kaempferia galanga* and *Elettariopsis curtisii* are used as flavours. In

Close-up of flower of *Alpinia galanga*.

Malaysia, the leaves of *Kaempferia galanga* (called *cekur*) is familiar in *perut ikan*, a local favourite dish. Yet another species, *Etlingera elatior* (locally known as *kantan*) is almost a compulsory ingredient of *laksa*, a favourite noodle dish in Peninsular Malaysia. *Kantan* is also used in the local dishes *nasi kerabu* or *nasi ulam* (of the East Coast of the Peninsula) and *laksa asam*. The young rhizomes of *Curcuma mangga*, *Boesenbergia rotunda*, and *Zingiber zerumbet* and young inflorescences of

Curcuma domestica and *Alpinia galanga* are also consumed as fresh vegetables or *ulam* (a term equivalent to salad) by village folk (Ibrahim, 1992).

Many studies and surveys have shown that at least more than ten cultivated species of Zingiberaceae have been frequently used in traditional medicine. Many of the medicinal gingers are used in traditional cures which are apparently associated with women-related ailments or illnesses, e.g., post-partum medicines for women during confinement. For such treatment, the species are consumed either on their own or in mixtures with other plant species either to regain their energy or for general health. The medicines are prepared in the form of decoctions, tonics or fresh rhizomes. Ginger species including zedoary *Curcuma zedoaria* (*temu kuning* in Malay), *C. mangga* (*temu pauh*), *C. aeruginosa* (*temu hitam*) and *Zingiber montanum* (*bonglai*) are used in food preparations for women in confinement after birth. Commercially, such medicines or tonics are available in the form of *jamu* which was originally a traditional Javanese herbal medicine from Indonesia. *Jamu* is made of a blend of several kinds of plants (Tilaar, Sangat-Roemantyo & Riswan, 1991). Examples of gingers used for the above purpose among others include, *Curcuma domestica*, *Boesenbergia rotunda*, *Zingiber officinale*, and *Kaempferia galanga*.

The rhizomes of *Zingiber officinale* and *Curcuma domestica* are frequently used as a carminative for relieving flatulence. The latter is also used as an anti-spasmodic in diarrhoea. Similarly *Curcuma xanthorhiza* (*temu lawak*), *Zingiber ottensii* (*lempoyang hitam* or *bonglai hitam*) and *Z. zerumbet* (*lempoyang*) have traditional roles in herbal medicine (Khaw, 1995). Besides these, *Globba* species are occasionally used in traditional medicine (Ibrahim, 1995).

Although gingers are better known in traditional medicine, and as spices, condiments or flavours, they are relatively new as ornamentals or landscape plants. The Zingiberaceae are mostly plants with showy inflorescences and often brightly coloured bracts and floral parts. In this they are second to none, not even to the orchids. For ornamental purposes, however, most of the species have one drawback: the flowers of most species are very shortlived, often lasting only a few hours. Many species do not thrive well outside the moist environment of the tropical rain forest. Therefore few species are commonly in cultivation. These include

Plants of *Kaempferia pulchra* develop equally attractive foliage and flowers.

Alpinia purpurata (Vieill) K. Schum., grown for its purplish red bracts (only the red-bracted form of *Alpinia purpurata* is common in Peninsular Malaysia and Singapore), *Hedychium coronarium*, *H. coccineum* Buch.-Ham., for their flowers produced in abundance, *Globba winitii* C.H. Wight with violet bracts, *Etlingera elatior* (Jack) K. Schum., mainly for its compact red inflorescence coloured by numerous bracts, *Curcuma roscoeana* Wall., and *Curcuma alismatifolia* Gagnep. (Siamese Tulip) both for their coma, the large coloured terminal sterile bracts. Some species of the Himalayan genus *Roscoea* have found their way into temperate gardens in Europe and North America, but they will not thrive in the Malaysian climate. Interestingly *Kaempferia pulchra*, a wild species discovered in the limestone hills of Langkawi Islands, in the far northwest of Peninsular Malaysia, has been successfully brought to cultivation growing in pots, on the ground in shaded areas or between crevices of rock gardens.

Farms in Australia and Costa Rica for instance are selling the inflorescences of *Etlingera elatior* (the torch ginger), in shades of pink, deep red and purplish black, as cut flowers. The inflorescence of *Zingiber spectabile* which changes its bract colour as it matures and has a shelf life of about two weeks, is also fast becoming popular as a cut flower.

4
Plant Structure in the Ginger family

Gingers vary in height and size from gigantic erect leafy shoots which can sometimes grow over 8 m high as in some species from the Alpinieae (e.g., *kantan* or *Etlingera elatior*). Others can be as small as 10 cm or less or even almost prostrate near the ground as in *Kaempferia galanga* (*cekur*) and *Camptandra parvula* (tribe Hedychieae).

All species are perennial herbs where the rhizome (the general term for an underground stem), which may be short or long, may be aboveground or subterranean. Each rhizome typically turns upwards, transformed into an erect leafy shoot, but the buds found near its apex can form new rhizomes which behave in the same way. These interconnected rhizomes form a series known as a sympodium (in Greek meaning "jointed feet"). Single plants will eventually form large clumps due to the spreading habit of the rhizomes. The clump size will depend on whether the rhizomes are of short or long elements and also the environmental factors. In some species (such as *Hornstedtia*) the rhizome is raised above the soil on stilt roots. In both the roots and rhizomes, oil cells containing aromatic compounds give a spicy smell when bruised.

The climate of Peninsular Malaysia allows continuous growth of the rhizomes throughout the year although more vigorous growth may be apparent during the rainy season. Nevertheless, in some genera like *Curcuma*, *Kaempferia*, *Zingiber* and *Globba*, in which cases the rhizomes are more or less fleshy, the rhizomes remain dormant and lose their leaves at the same time, for a certain period of time especially during the dry season, and then almost burst into life just after the rain. Dormancy in *Curcuma* and *Globba* is drought-induced.

As mentioned earlier, the rhizomes are usually aromatic but in some species it is the roots which are strongly fragrant. In any case, it is also the rhizomes which are frequently consumed or utilised in mixtures for food and medicine.

A real stem is present in many species, but it is usually short, while a pseudostem (or false stem) formed by the leaf sheaths may reach as high as 8 m. The base of the leafy shoot is covered by several leaf sheaths without properly developed blades, followed by the normal leaves arranged in two rows. Each leaf sheath is terminated at its apex by a membraneous ligule, and continues upward as a narrowed stalk-like portion, and then the leaf blade. The leaf blade may be symmetric or asymmetric (unequal) at its base. Asymmetric leaf bases are characteristic of the Hedychieae group of species. Leaves vary from smooth (glabrous) to slightly hairy or even (very rarely) densely hairy. The leaves of the different genera vary in size from just a few centimetres long (some species in the Hedychieae group) to as long as a metre (as seen in some species in the Alpinieae group).

Although typically, leaves exhibit shades of light to dark green in colour, some are flushed with purple underneath and some exhibit striking streaks of silver on the upper surface (*Kaempferia*, *Boesenbergia*, *Scaphochlamys*) rendering them commercially popular as foliage plants. Several species and cultivars of the ornamental alpinias have developed variegations on their leaves creating an exotic foliage display.

The stalk-like part that joins the blade to the leaf sheath proper has been called the "petiole". In cases where there is more than one leaf, a gradation in size, shape and length of petiole occurs from the lower (or outer) to the upper (or inner) leaves. In the genera *Curcuma*, *Scaphochlamys*, *Boesenbergia* and *Elettariopsis*, the petiole can be very long in relation to the blade and is sometimes channelled. Interestingly, the genus *Zingiber* has a pulvinus-like (swollen) petiole similar to what is found in Marantaceae, which distinguishes it from all the other genera in the family.

The inflorescence may terminate the leafy shoot (i.e., is "terminal") or it may be borne on a leafless shoot on the rhizome (i.e., it is "radical") near the leafy shoot or at some distance from that. If terminal, there will be a long peduncle (main stalk) with bracts in the distal part. The bracts each subtend a "partial inflorescence" with many or few flowers; in rare cases, in the genus *Alpinia*, it is reduced to one flower, in which case the inflorescence resembles a spike. In species with radical inflorescences, these may be partly covered in the ground or borne on a peduncle. Extremely rare is the case when both terminal and radical inflorescences occur on the same plant. Such a phenomenon was seen in *Zingiber puberulum* in the wild.

Several species of the genera *Amomum*, *Elettaria*, *Elettariopsis*, *Geocharis*, *Geostachys* and *Zingiber* are rather unique in having their inflorescence prostrate or nearly so. Even more remarkable is the inflorescence of the genus *Plagiostachys* which appears to be breaking through the side of the sheaths of the leafy stem. Curiously, a similar inflorescence structure occurs in *Alpinia havilandii* of Malaysian Borneo.

The inflorescence in Zingiberaceae is generally described as a spike or raceme. Only in the genus *Plagiostachys* and few species of *Alpinia* is there a paniculated inflorescence. When the inflorescence bracts are tightly overlapping, the inflorescence appears cone-like. In principle, the inflorescence consists of a main axis, alternatively called a rachis, bearing primary bracts. The bract is usually open to the base and primarily boat-shaped. The inflorescence bracts are usually spirally arranged when numerous; rarely they are few, solitary and cup-shaped (e.g., *Camptandra*) or lacking. In *Curcuma*, for instance, the bracts unite laterally to form pouches but in *Boesenbergia* the bracts are unusually distichous. In some genera (*Curcuma*, *Etlingera*, *Zingiber*) the primary bracts may be large, showy and brilliantly coloured in hues of pink, yellow, orange, purple or white. These features have been frequently exploited for commercial purposes.

All bracts other than those that arise on the main axis or branches thereof, may be called secondary bracts or bracteoles. The bracteole is almost always dull and inconspicuous. Bracteoles are present in most groups.

The flowers are highly specialized and bisexual. They range from tiny forms as in globbas to fairly large types as in alpinias. Ginger flowers are

Bracts and Bracteoles

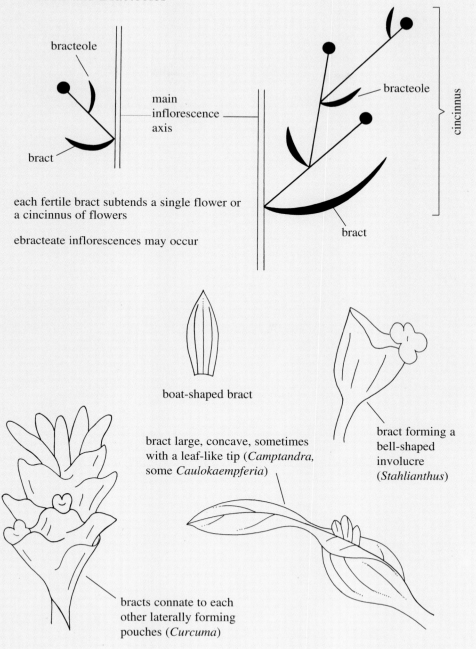

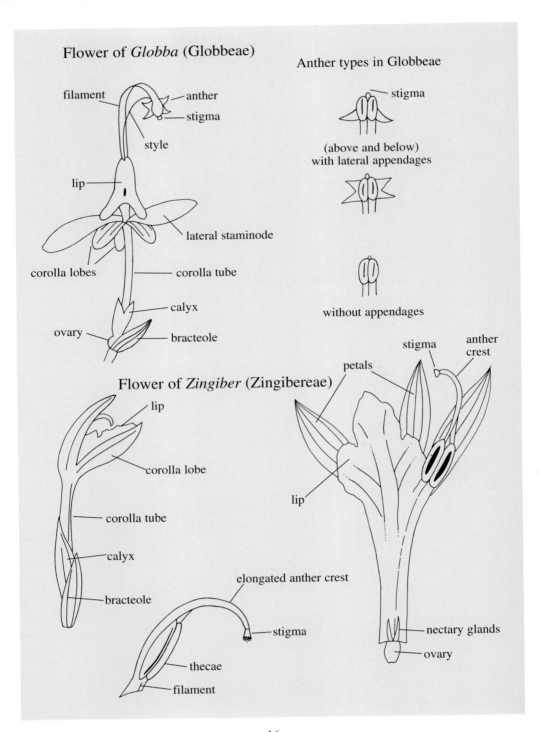

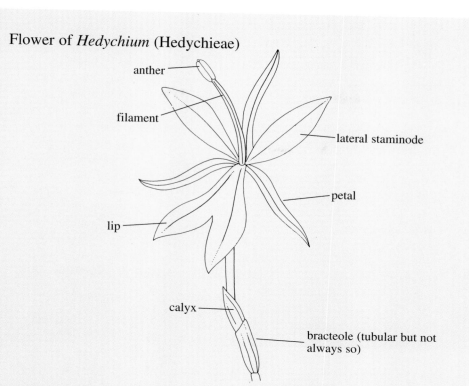

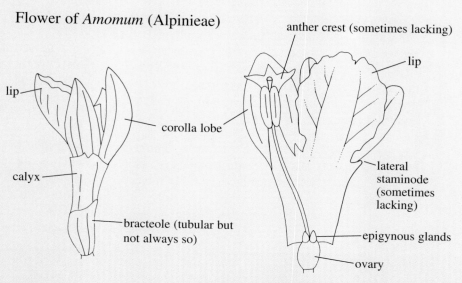

(Drawn by Yong Ket Hyun after Rosemary Smith)

fragile and ephemeral, emerging from base to apex of the inflorescence in most genera, and the other way around in *Boesenbergia* and *Haplochrema* (Hedychieae).

The flower has an inferior ovary, this may be unilocular with the ovules on three placentas along the wall or, more commonly, 3-locular with axile placenta. On top of the ovary there are two so-called stylodial nectiferous glands. The calyx is tubular with three, rarely two, teeth. The corolla consists of a narrow corolla tube and three subequal corolla lobes, of which the upper (dorsal) one usually is of a slightly different shape. Superficially, the flower of Zingiberaceae resembles that of orchids in having a striking lip or labellum (except the much reduced lip of *Burbidgea*, a genus endemic to Borneo). Botanically, the lip in gingers is actually formed from 2–3 modified sterile stamens in contrast to the orchid lip, which is a true petal.

The two lateral stamens are transformed into petal-like (petaloid) structures in the Globbeae and Hedychieae or into two small teeth (or wanting) in Alpinieae and Zingibereae. The single fertile stamen has a short or long filament terminated by the anther, which in most species opens by longitudinal slits, and more rarely by terminal pores. The anther may be provided with various appendages. In the genus *Globba*, these appendages are triangular and positioned along the sides of the anther, one or two on each side. In several genera, e.g., *Kaempferia* and *Boesenbergia*, the anther connective is prolonged into a terminal crest. In the genus *Curcuma*, the anther is provided with basal spurs. The style is thin and long and placed in a furrow in the filament and the anther, and protrudes above the anther.

The fruit is a dry or fleshy capsule that may dehisce by three longitudinal slits or it may be indehiscent. In texture it is mostly smooth, slightly ridged or spiny, sometimes hairy and rarely ribbed. In some species of *Etlingera*, the single fruits are fused into a large, composite, fleshy fruit. They are generally green when young but turn orange, brown, maroon or even black on ripening. The seeds are dark brownish to black, provided with a more-or-less lacerate red, orange or whitish aril or fleshy jacket. For many species, the fruit and seeds are not yet scientifically documented. The seeds can be strongly aromatic, as in *Elettaria cardamomum*, the cardamom of commerce.

5
Pollination and Seed Dispersal

The biology of the Zingiberaceae is poorly known. Pollination has only been observed in few species, but butterflies and moths seem to play a major role. Ants and bees have been seen to visit flowers of several *Amomum* and *Alpinia* species, and may be pollinating agents. Unlike in Heliconiaceae, birds as pollinators are rare in Zingiberaceae except that of *Etlingera elatior* as reported by Classen (1987).

A Skipper butterfly pollinating a flower of *Alpinia mutica* at the Rimba Ilmu, the University of Malaya's botanical garden in Kuala Lumpur.

In the genus *Hedychium*, where several flowers are open at the same time, butterflies are sometimes trapped in the long narrow corolla tube. In their effort to free themselves they pollinate several neighbouring flowers. Some *Hedychium* species are pollinated by moths.

The dispersal of seeds is even less known. The open capsules of *Hedychium* are strongly orange-coloured on the inner side, and the red arillate seeds, fused into a single mass in the open fruit, most certainly will attract birds, although bird dispersal has never actually been documented. Many species, fruiting near the ground, have a white, lacerate seed aril, and may be ant-dispersed. The large, fleshy syncarps of some *Etlingera* species are undoubtedly eaten and dispersed by animals. In the genus *Caulokaempferia* found in southern Thailand, not yet recorded from Malaysia, seed dispersal by rain splash or splash from small streams has been observed. The small plants often grow on rocks near streams and produce numerous minute seeds in an apically open capsule.

(Left) A Lycaenid butterfly helps itself to nectar from the flowering head of the Torch Ginger, *Etlingera elatior*. (Right) A bee (*Amegilla* sp.) hovers before newly open flowers of *Globba schomburgkii*.

6
A Brief History of Research on Malayan Gingers

The gingers have been recognised as a group from the earliest days of formal botany. In the first volume of his *Species Plantarum*, Linnaeus listed the Zingiberaceae under "Classis 1, Monandria Monogynia" on the very first page. He enumerated three species of *Canna*, four species of *Amomum*, one of *Costus*, one of *Alpinia*, one of *Maranta*, two of *Curcuma* and two of *Kaempferia*. Eight of these species are said to be from "India". None of the types could be referred to specimens from Peninsular Malaysia.

The first botanist to collect gingers from the Malay Peninsula was the Danish scientist J.G. Koenig, one of the greatest plant collectors in the 18th century, who during his stay on the Siamese island of Phuket in 1779 collected numerous Zingiberaceae. These species were described in Retzius' *Observationes* in 1883. Koenig's plant specimens were for a long time regarded as lost but have recently been 'rediscovered' in the Botanical Museum of Copenhagen.

Koenig's descriptions are among the finest in the classical literature. Another Danish botanist, who brought the study of Malaysian and Indian Zingiberaceae a great step forward, is Nathaniel Wallich (born in Copenhagen as Nathan Wulff). Wallich succeeded William Roxburgh as director of the Botanic Garden in Calcutta and published the classic *Plantae Asiaticae Rariores* (The Rare Plants of Asia). His type specimens are in Kew and in Copenhagen. In the standard work, the *Flora of British India* by J.D. Hooker, which documented the rich flora of the Indian area, including what was then known about the Malay Peninsula, J.G. Baker provided an account of the Zingiberaceae (published in volume 6) in 1892. The main set of his types are in the Kew herbarium, but many are also deposited in the Singapore herbarium.

The first work dealing solely with the Zingiberaceae is H.N. Ridley's *The Scitamineæ of the Malay Peninsula* in the Journal of the Straits

Ridley in his office, photo taken by Charles de Alwis around 1900. (Reproduced with kind permission of the Singapore Botanic Garden)

Branch of the Royal Asiatic Society, published in June 1899. It is tempting to quote from this:

The traveller in the forests of the Peninsula can hardly fail to notice the beauty of many of our wild gingers (Scitamineæ) and would be surprised to find how much this interesting group of plants has been neglected by botanists, for though many have received names, but few have been completely described... and further Many descriptions have been made from badly dried specimens, and unless special care is taken these plants do not preserve well.

All this and more from Ridley's introduction would still be correct if they were written today. For its time, it was an excellent work in which he described 53 new species.

The Dutch botanist T. Valeton worked first as bacteriologist at the Java Sugar Experiment Station in East Java but later, from 1892 to 1913, he became scientist at the Buitenzorg (now Bogor) Botanic Garden. He worked on several families but particularly on Rubiaceae and Zingiberaceae. His thorough descriptions of species from Java, Borneo and the Philippines were mostly accompanied by fine drawings.

R.E. Holttum. (Reproduced with kind permission of the Singapore Botanic Garden)

The great German botanist K. Schumann published his account of the *Zingiberaceae of Malaisia and Papuasia* in 1899 and completed his monograph on the Zingiberaceae for Engler's monumental *Pflanzenreich* in 1904. Critical remarks have been raised on this work, not least by Holttum, but with the knowledge available at that time, it was a great leap forward and is still in many ways a fundamental work. The next work to make a definite contribution to the study of Peninsular Malaysian gingers was R.E. Holttum's *The Zingiberaceae of the Malay Peninsula*, published in the Gardens' Bulletin, Singapore, Volume 13, in 1950. In this work, the author carried out a thorough analysis of the inflorescence and flower structure of the genera of Zingiberaceae and revised the whole group for what is today Peninsular Malaysia.

Much material was not available to Holttum during his work and after completing it he turned his interest to the Pteridophytes. Therefore, many problems were not solved and left for the future. It is hoped that the coming revision of the family for the *Flora Malesiana* (which attempts to document the flora of the Malesian botanical region) will not be too far away. Anyway, more botanists than ever before are working on the Zingiberaceae of Southeast Asia and today, more collectors have found fascination in the beauty of our wild gingers, to repeat Ridley's sentiment.

Major reference works dealing with the Zingiberaceae in tropical Asia include Baker (1892), Gagnepain (1908), Holttum (1950), Loesener (1930), Ridley (1899, 1925), Roscoe (1824–28), Schumann (1899, 1904), Wu (1981), and Wu *et al*. (1996).

7
The Tribes and Genera of the Zingiberaceae in Peninsular Malaysia and Singapore

As mentioned earlier, our knowledge is still poor for many groups within this large tropical family. In all areas where intensive studies have been undertaken during recent years, numerous new species, even new genera, have been discovered. We are convinced that less than 90% of the species in Southeast Asia have been discovered. Recent intensive collections in, e.g., Thailand, Sabah, Sarawak and Brunei have revealed a richness in the Zingiberaceae that astonishes most specialists. Areas that are still badly in need of collection include Sumatra, Kalimantan and Sulawesi.

Peninsular Malaysia and Singapore remain the most thoroughly studied areas in Malesia, but still new species may be found, e.g., in rain forest areas such as the Belum reserve. Holttum's important foundation work on the Zingiberaceae of the Malay Peninsula, published in 1950, amply discusses how little we know about the delimitation of many species and how many species for which the fruits are unknown.

Among Peninsular Malaysian Zingiberaceae, the highest diversity is recorded for the Alpinieae (9 genera, 84 species) followed by Hedychieae (7 genera, 52 species). The two tribes, Globbeae and Zingibereae, which are represented by one genus each in Peninsular Malaysia, are much less diverse with 15 species and 19 species, respectively. Of the 20 genera documented for Malaysia, only the genus *Haniffia* is endemic to Peninsular Malaysia. Despite the high diversity reported for Malaysia in general, and Peninsular Malaysia in particular, surprisingly only about 10% of these is known to the local people as common cultivated, wild or semi-wild species.

Gingers thrive in a wide range of habitats ranging from riverine to limestone rocks and from the lowlands to the upper montane regions. Most gingers are terrestrial, growing naturally in damp, humid, shady areas with good light but several native species can tolerate the full exposure of the sun. Epiphytic gingers are uncommon and of the two species of *Hedychium* which are usually epiphytic on tree trunks and branches; only *Hedychium longicornutum* has been frequently encountered.

Gingers are generally abundant in lowland to hill forests, notably between 200 m and 500 m above sea level. Gingers are less profuse in higher altitudes and rather scarce on very high mountains. On small isolated islands which are relatively dry, the diversity of gingers is quite low and sometimes they are even absent. Some species inhabit secondary forests, open places such as along road-sides, forest gaps, riverbanks and swampy vegetation. In Peninsular Malaysia only one species of *Zingiber* is known from a peat swamp forest. Some species of *Etlingera* for instance, rapidly inhabit disturbed secondary forests or newly opened areas and subsequently spread like weeds. Some of these species however are useful indicators of disturbed habitats.

Among the wild gingers, several species are widespread, including *Alpinia javanica*, *Amomum biflorum*, *Amomum uliginosum*, *Etlingera punicea*, *Globba pendula* and, to a lesser extent, *Camptandra parvula*. Some other species are apparently localised, in particular, several species of *Scaphochlamys* (most are lowland species) and *Geostachys* (almost all are mountain species).

An overview of the tribes and genera

Gingers are conventionally classified into distinct genera (each consisting usually of several to many species) and the genera also show affinity as groups recognised as tribes. A brief survey of these is given.

Key to the tribes

1. Small plants up to 0.5–1 m with inflorescence terminal on leafy shoots, flowers with a long-exserted stamen and a bifid lip attached to the filament. Flowers orange-yellow or violet-white (page 27–30) .... 1. Globbeae (*Globba*)
   Not this combination 2

2. Medium-sized plants from 50 cm upwards with stout rhizome, radical inflorescence (i.e., terminal on specialized leafless flowering shoots arising directly from the rhizome), and flowers of various colours with an undivided lip and anther with a long, curved terminal anther appendage (page 31–36) .... 2. Zingibereae (*Zingiber*)
   Not this combination 3

3. Flowers with well-developed lateral staminodes (page 40–55) .... 3. Hedychieae
   Flowers with lateral staminodes wanting or reduced to small teeth (page 57–92) .... 4. Alpinieae

1. Globbeae

Globba

Slender herbs, rarely above 50 cm, flowering from the top of the leafy shoots. Flowers yellow, white or violet, in some species with overlapping bracts; in others, the inflorescence has widely spaced bracts. In several species bulbils replace the flowers. Bulbils may also occur in the leaf axils. The flower is very characteristic with its lip joined to the stamen and the long-exserted, curved stamen. In all Peninsular Malaysian species, the anther is spurred with either one or two triangular appendages along either margin: no other genus has this character. The species are mostly found in shady forest. Species determination is often difficult due to hybridization between the species, especially within the *Globba cernua* complex. Some species have ornamental value.

Globba corneri.

Globba variabilis spp. *pusilla*, with dark green, velvety leaves and orange flowers.

(Right) *Globba nawawii*, with delicate white bracts and orange flowers. This is a rare species so far known only from the Peninsular Malaysian state of Terengganu.

(Below) Somewhat a misnomer, *Globba unifolia* was thought to be a single-leaved species from its type specimen, but in fact typically bears two to four leaves per shoot. The short, stiffy deflexed inflorescence bears rose-pink to purplish red bracts and orange-coloured flowers.

(Left) Flowers of *Globba pendula*, one of the commonest and most variable species of the genus throughout Malaysia, Singapore, southern Thailand and Indonesia.

A colony of *Globba fragilis* clothes a boulder, thriving only on a thin layer of organic matter.

(Right) *Globba schomburgkii*, with flowers in the distal part of the inflorescence and bulbils along the basal part.

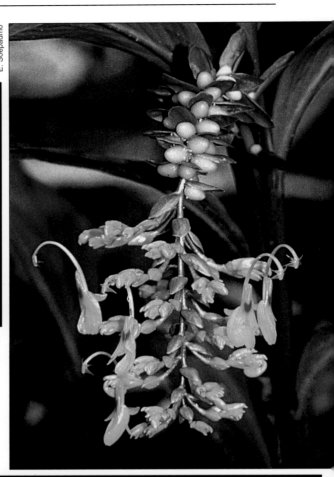

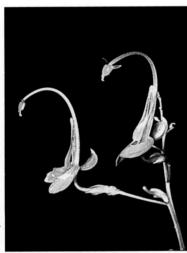

(Above) *Globba leucantha*: there is a great range in flower colour from white with lilac markings to almost pure violet or yellowish white.

(Below) *Globba patens*, one of the more easily recognized species of *Globba* with orange-coloured flowers crowded on the short rachis.

2. Zingibereae

Zingiber

Medium-sized herbs with long, creeping, stout rhizomes. The inflorescence arises on a separate shoot directly from the rhizome. The peduncle (main stalk of the inflorescence) may be short or long. The inflorescence has closely overlapping bracts, or the bracts form an open pouch in which the flowers occur, one to each bract, the whole inflorescence often with a cone-like appearance; in many species the bracts are green when young, turning red in the fruiting stage. The flowers are usually very

(Above) A *Zingiber zerumbet* inflorescence with white flowers (lemon yellow lip).

(Right) The delicate yellow flowers of *Zingiber zerumbet*, a widely cultivated medicinal plant which is probably native to south India and Sri Langka.

ephemeral, only lasting a few hours. The lip is 3-lobed. The unique character is the stamen that is provided with a long, curved beak- or horn-like appendage. The name *Zingiber* actually comes from a sanskrit word for a bull's horn. The true ginger, *Z. officinale*, belongs to this genus.

There seems to be much confusion, particularly in the *Zingiber gracile* complex. Several varieties which are difficult to discriminate are also encountered in the *Zingiber puberulum* group.

Zingiber officinale, commonly known as *halia* in Malaysia, has two races known in Peninsular Malaysia as *halia bara* and *halia padi*. *Halia bara* differs from the typical halia in its much smaller, red rhizomes, which have also more pungent tissues. *Halia padi* is very similar to *halia bara* in the size of its rhizomes, although the yellow colour of the typical ginger is retained. In the colour and structure of the inflorescence, these races are indistinguishable from the typical ginger, or *halia*. All these types of ginger flower only very infrequently in Malaysia.

(Above) The club-like, chestnut-brown apical part of the inflorescence of *Zingiber multibracteatum*, with elegant white and pink flowers.

(Right) A smallish forest ginger of the lowlands, *Zingiber citrinum*.

Zingibereae: Zingiber

(Above) Old flowering spikes protruding above the ground at the base of a clump of *Zingiber puberulum*.

(Above) Closer view of a *Zingiber puberulum* flowering spike with a freshly open flower.

(Left) *Zingiber ottensii*.

(Above) An attractive, brightly coloured orange-red flowering spike of *Zingiber kunstleri*, an uncommon species in the Peninsula, at Kuala Koh in Kelantan.

(Right) *Zingiber griffithii*, closely related to *Z. puberulum*, produces a fusiform inflorescence with densely imbricate bracts that eventually develop a pink to reddish or yellowish colour.

(Far right). *Zingiber fraseri* produces an erect inflorescence, the bracts maroon as they mature, with pale yellow flowers.

Zingibereae: Zingiber

The pouch-like, yellow inflorescence bracts with incurved margins and the dark purple flowers with many small dots help make *Zingiber spectabile* distinctive.

Zingiber spectabile, habit.

Zingiber sulphureum, in cultivation in the Royal Botanic Gardens, Kew.

3. Hedychieae

Key to the genera

1. Bracts joined along the margins in their lower part, forming pouches in which the flowers develop. Terminal part of inflorescence with a coma .... 3. *Curcuma*
   Bracts not joined, never forming a coma 2

2. Filament long-exserted .... 5. *Hedychium*
   Filament not exserted, shorter than lip 3

3. Several flowers to each bract 4
   One flower to each bract 5

4. Inflorescence enclosed in one main concave bract .... 2. *Camptandra*
   Several bracts forming the inflorescence .... 7. *Scaphochlamys*

5. Lip bilobed to the base, resembling a 4-merous flower, violet or white .... 6. *Kaempferia*
   Lip entire or shallowly bilobed 6

6. Inflorescence terminal on the leafy shoot .... 1. *Boesenbergia*
   Inflorescence on a separate shoot .... 4. *Haniffia*

1. Boesenbergia

Small forest plants with shoots consisting of 1–few leaves, often with purple sheaths. The inflorescence arises, in all Peninsular Malaysian species, between the leaves and is more-or-less enclosed by the leaf sheaths. The bracts are distichous, each subtending one flower. The flowering begins from the top of the inflorescence. All floral parts are extremely delicate and shortlived. In most species the lip is prominent, saccate and whitish or yellowish, sometimes flushed red. In *B. rotunda* it is uniformly pinkish and in *B. curtisii* it is small, flat and yellow with a purplish patch. One species, *B. rotunda* (formerly *B. pandurata*), not native to Peninsular Malaysia, is widely cultivated as a spice and for traditional medicine.

(Right) *Boesenbergia plicata* (fomerly *B. plicata* var. *lurida*) with red flowers, is common in southern Thailand and the islands of Langkawi, Terutau and Tioman.

(Above) Flower of *Boesenbergia plicata*.

(Opposite page) *Boesenbergia plicata*. (Inset: flower)

Hedychieae: Boesenbergia

(Right) *Boesenbergia curtisii*, a limestone species found from southern Thailand to northern Peninsular Malaysia.

(Below) A widely cultivated species for its rhizomes that are used in Malaysian and Thai cooking, *Boesenbergia rotunda* is probably indigenous to India and south China (Yunnan). The inflorescence is covered by the leaf sheaths and the flowers, uniformly pink, appear one by one.

2. *Camptandra*

Small herbs with a leafy stem. Leaf blade asymmetric with strongly marked veins. The inflorescence is terminal on the leafy stem and easily recognizable by the one large, concave green bract subtending a small inflorescence with few flowers. Flowers small, white to pale violet, with a broad, bilobed lip. The stamen is very short. There are only three species in the genus, apparently entirely restricted to Peninsular Malaysia and Borneo. The small plants are easily overlooked, but *C. parvula* is quite common on riverbank rocks and near waterfalls, from the lowlands to the mountains. *C. latifolia* and *C. ovata* are both species of higher elevations above 1000 m.

Camptandra ovata is found at elevations of 1000 m and above, in moist shady habitats.

3. Curcuma

Usually small to medium-sized herbs with fleshy rhizomes and tuber-bearing roots. The inflorescence is formed either between the leaves or on a separate shoot with a short scape. The inflorescence is composed of broad bracts joined to each other to form closed pouches in which small partial inflorescences develop. The bracts may be green or coloured and the whole inflorescence is often crowned by a rosette of coloured sterile bracts, called a coma. The flowers are usually of a colour different from the bracts. The upper corolla lobe and the staminodes overlap, forming a hooded structure, while the lip is rather short with upcurved margins. The stamen is short, situated under the hood, and characteristic by its anther provided with two curved spurs at its base, and a short crest at its apex. It is a large genus with only few indigenous species in Peninsular Malaysia, while 5–6 species are cultivated for medicinal use and for flavouring food (turmeric). The genus name is derived from the Arabic *kurkum*, referring to the yellow colour of the turmeric rhizome.

(Right) *Curcuma* cf. *aurantiaca*, common in Java but not in Peninsular Malaysia and Singapore, has potential as a cut-flower plant. Note the pink coma formed by the upper inflorescence bracts and green lower bracts.

Khaw Siok Hooi

Yong Hoi Sen

(Above) *Curcuma domestica*, the turmeric, one of the oldest spice plants known, is called the *kunyit* in Malay.

Hedychieae: Curcuma

Curcuma aeruginosa, a species widely distributed throughout Southeast Asia, and sometimes cultivated, produces rhizomes used medicinally.

4. Haniffia

Small herbs with leafy stems. Inflorescence on separate side shoots from the base of the leafy shoot, with slender rachis. Flowers without bracteoles, staminodes narrow, lip not deeply bilobed, anther lacking in basal appendages.

This genus was named in honour of Mohamed Haniff, an active field worker of the Penang Botanic Gardens during the time of the Straits Settlements, who made many collections of Peninsular Malaysian plants. Only one species, *H. cyanescens*, has been described for Peninsular Malaysia. This genus is believed to be related to *Roscoea*.

5. Hedychium

Medium-sized gingers, some species epiphytic. Rhizome stout, fleshy. Inflorescence terminal on the leafy stem, often curved at base. Bracts broad and closely overlapping forming a more-or-less cone-like inflorescence. Flowers white, yellowish, orange or red. The corolla tube is slender, long-exserted from the bracts, the lobes long and narrow, usually

Hedychium collinum, endemic to Peninsular Malaysia.

reflexed. Staminodes broader, conspicuously coloured. The lip has a narrow basal part and a broad, deeply divided distal blade. The stamen is very long, often curved, terminated by the anther which is divided at its base. The long-exserted stamen is unique among the gingers of the Malay Peninsula. Unlike most other gingers, several flowers develop at the same time. They are pollinated by butterflies, which are often trapped by the narrow corolla tube. The name *Hedychium* comes from the Greek words, *hedys* (meaning sweet) and *chion* (meaning snow), referring to the fragrant white flowers of the well-known *H. coronarium*.

(Above) *Hedychium longicornutum*: the flowers are yellowish to reddish orange and have long, creamy yellow to pinkish filaments.

(Right) The epiphytic *Hedychium longicornutum* clings on to its support with a dense mass of fleshy roots. It is quite common in lowland and freshwater swamp forests in the Malay Peninsula but has not been recorded for Singapore.

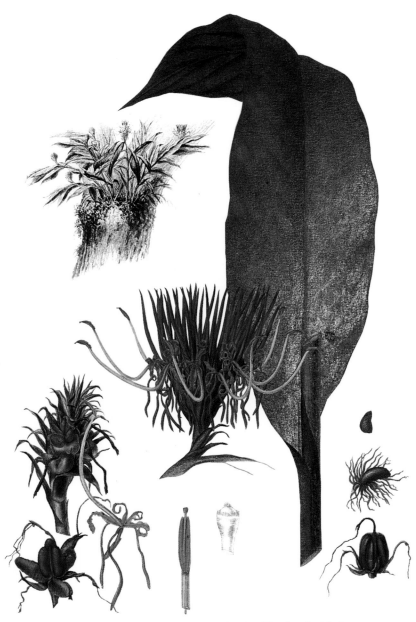

(Above) *Hedychium longicornutum*. (Painting by Charles de Alwis. Reproduced with kind permission of the Singapore Botanic Garden)

(Opposite) *Hedychium paludosum*, apparently restricted to the Cameron Highlands area in Peninsular Malaysia.

(Above) Native to Myanmar, northeast India and southern China, *Hedychium coronarium* is cultivated as an ornamental throughout the tropics.

(Right) *Hedychium coronarium* inflorescences are spindle-shaped and the fragrant, white flowers emerge from the overlapping green bracts.

6. Kaempferia

Small herbs with short rhizome and tuberous roots. In the indigenous species, the flowers arise in the midst of a few leaves, while in the introduced species *K. rotunda*, the inflorescence appears before the leaves. The inflorescence is totally enclosed in the leaf sheaths. The genus is easily recognized as the flowers appears to consist of four lobes, surrounded by three thin, narrow corolla lobes. In each flower, two broad staminodes and a twice-as-broad labellum divided almost to the base catch the eye. Only two species are indigenous. *K. galanga* is probably introduced from India. It is still used in traditional medicine. The name *Kaempferia* commemorates Engelbert Kaempfer, a German physician and botanist in the 17th century.

(Above) Flower of *Kaempferia rotunda*.

(Left) *Kaempferia rotunda*, like *Kaempferia galanga*, is introduced into cultivation in Malaysia and Singapore.

(Above) Plants of *Kaempferia pulchra* develop equally attractive foliage and flowers.
(Below) Flower of *Kaempferia pulchra*.

(Right). Flower of *Kaempferia galanga*, the *cekur*, the leaves of which are used in food flavouring.

7. Scaphochlamys

Small plants on the forest floor, with creeping, thin rhizomes, in some species supported by stilt roots above the ground. Leaves one to few. Inflorescence terminal. The inflorescence is carried by a shorter or longer scape (or peduncle, i.e. the main inflorescence stalk), or the flowers appear just above the leaf sheaths. The inflorescence bears some resemblance to that of the genus *Curcuma*, but the bracts are free, each subtending one to several flowers. The flowers are white or creamy coloured and often with a median yellow band bordered with pink-purplish streaks or spots, only slightly emerging from the bracts. The labellum is flat, obovate, and more or less deeply divided. The genus is polymorphic and mainly concentrated in the Malay Peninsula with about 20 species and many are local endemics.

(Above) *Scaphochlamys sub-biloba*, endemic to Tioman Island. (Left) *Scaphochlamys sub-biloba*, flower.

Scaphochlamys kunstleri, white-flower form with a yellow median band and pink stripes on both sides of the base of the lip.

Hedychieae: Scaphochlamys

(Left) *Scaphochlamys concinna.*

(Above) *Scaphochlamys longifolia* has a short inflorescence with green bracts, and flowers that are white and pink-tinged, the lip itself decorated with reddish-pink streaks.

4. Alpinieae

Key to the genera

1. Inflorescence breaking through the leaf sheaths, appearing lateral on the shoot .... 9. *Plagiostachys*
   Inflorescence distinctly terminal on the leafy shoot, or on a separate side shoot 2

2. Inflorescence distinctly terminal .... 1. *Alpinia*
   Inflorescence on a separate side shoot 3

3. Inflorescence dense, cone-like, bracts overlapping 4
   Inflorescence lax, bracts not overlapping 7

4. Inflorescence surrounded by several sterile, coloured bracts larger than the floral bracts 5
   Inflorecence not surrounded by coloured sterile bracts 6

5. Inflorescence not spindle-shaped, lip stiffly incurved after flowering, stamen shorter than lip .... 5. *Etlingera*
   Inflorescence spindle-shaped with striate or reticulate nervation, lip not stiffly incurved, stamen usually as long as lip ... 8. *Hornstedtia*

6. Inflorescence cone-like, bracteoles tubular .... 2. *Amomum*
   Inflorescence dense, not cone-like, bracteoles not tubular .... *Elettariopsis triloba*

7. Inflorescence erect or decurved, not prostrate or partly buried in the ground 8
   Inflorescence prostrate, in some species underground except for the flowers (from the corolla tube upwards) 9

8. Lip split to the base .... 6. *Geocharis*
   Lip not split to the base .... 7. *Geostachys*

9. Flowers several together (in cincinni) .... 6. *Elettaria*
   Flowers solitary on a creeping axis .... 4. *Elettariopsis*

1. *Alpinia*
(including the segregate genera *Catimbium*,
Cenolophon and *Languas*)

The largest genus in the family, with over 250 species in tropical Asia. They are medium-sized to large forest plants, some species reaching a height of over 3 m. It is the only genus in the Alpinieae that has a terminal inflorescence on the leafy shoot. The flowers are yellowish-green to creamy coloured or red, usually conspicuous. The staminodes are reduced to large teeth (several mm long) at the base of the lip. The lip is more-or-less saccate and not divided, if pale coloured often with yellow blotches or red lines. The capsules are smooth, spherical or

Yellow and white blooms of *Alpinia javanica*.

ellipsoid. Holttum, in his treatment of the gingers of the Malay Peninsula, advocated splitting the genus into the four segregate genera. Recent studies on morphological variation in the group from the whole of its distribution area does not support this. There is one economically important species, *A. galanga*, used medicinally and as a spice. This species is not native to the Malay Peninsula, but is widely cultivated. The genus commemorates the 16th-century Italian botanist, Prospero Alpinio.

(Above) Fruits of *Alpinia mutica*, like the flowers, are produced almost continuously. The seeds are aromatic.

(Right) *Alpinia murdochii*, endemic to Peninsular Malaysia, was described from Fraser's Hill by Ridley.

Alpinia pahangensis.

Alpinia purpurata, an introduced species in Peninsular Malaysia and Singapore, includes a form with pink inflorescence bracts.

Alpinieae: Alpinia

Alpinia malaccensis. (Painting by Charles de Alwis. Reproduced with kind permission of the Singapore Botanic Garden)

Alpinia nutans, later recognized as *A. zurumbet*. (Painting by Charles de Alwis. Reproduced with kind permission of the Singapore Botanic Garden)

Alpinieae: Alpinia

(Left) *Alpinia scabra*, widely distributed from Peninsular Malaysia through Sumatra to Java, in flower. (Right) *Alpinia scabra*, infructescence.

The flowers of *Alpinia*, often incredibly beautiful, will need to be more closely studied to help elucidate correct specific identity and relationship.

Alpinia petiolata, a species relatively common in the montane forest in Peninsular Malaysia.

Alpinieae: Alpinia

(Above) *Alpinia oxymitra*, with characteristically slender leaves and narrow, ribbed fruits ripening yellow-brown, grown in the Rimba Ilmu botanical garden of the University of Malaya in Kuala Lumpur.

(Left) The smallest flowers among Peninsular Malaysian species of *Alpinia* are found in *A. conchigera*.

(Right) Inflorescence of *Alpinia galanga*.

Alpinia galanga, introduced and widely cultivated in Peninsular Malaysia for use in folk medicine and as a spice.

2. Amomum

The second largest genus in the family. The vegetative parts have often a strong smell of cardamom when crushed; the leafy shoots are often tall; the inflorescence is on a separate side shoot with loosely overlapping bracts. In some species it may be partly covered in the soil, on others on a short peduncle. The flowers are yellow or straw-coloured with red stripes. The labellum is slightly longer than the corolla lobes. Many species are easily separated on fruit characters as the capsules, which may be dry or fleshy, have a characteristic surface covered by spines or ridges or variously sculptured; in some species the capsules are smooth. The genus name is derived from the Greek words *a* (without) and *momos* (harm), referring to the antidote property of an unknown spice plant.

(Above) *Amomum spiceum*, open flower on an inflorescence.

(Left). Inflorescences from the base of an *Amomum spiceum* clump.

(Above) *Amomum testaceum*, flower.

(Right) The rather large inflorescence of *Amomum testaceum*, up to 15 cm long, is covered with papery, straw-coloured bracts. Perhaps identical to the Siamese cardamom, *A. krervanh*, this species requires more careful study to establish its correct identity.

Alpinieae: Amomum

(Above) One of the larger species of the genus, *Amomum ochreum* is conspicuous in flower by its pinkish-red petals.

(Left) *Amomum uliginosum* was originally described by Koenig from Phuket island (Thailand), which was then sometimes called Junk-Ceylon, a name often mistaken even by botanists for Sri Lanka.

(Above) The orange-yellow flowers of *Amomum aculeatum*.

(Above) *Amomum kapulaga*, flower.

(Right) *Amomum lappaceum*: an elongating, mud-covered, club-shaped flowering spike.

3. Elettaria

The genus is closely allied to *Amomum* but differs in its inflorescence structure. Tall herbs with stout rhizomes. Inflorescence from the base of the leafy shoot, prostrate, bearing two rows of sheaths. Bracts tubular; flowers short- or long-tubed; labellum with yellow median band and red stripes (similar to *Amomum*); filament short, anther longer than filament and crested. Fruit ellipsoid or globose, ridged or smooth. Apart from *Elettaria cardamomum* (the commercial cardamom), only one species, *E. longituba*, is known in Peninsular Malaysia. The name originated from a vernacular name used in Malabar (India).

4. Elettariopsis

Small plants with widely creeping, slender rhizomes. The leafy shoots, arising 10–20 cm apart, are rarely above 1–2 m with 1–8 petiolate leaves. Leaves often fairly large, in one species up to 1 m, but in the other species much smaller. The inflorescence is on a separate side shoot at the base of the leafy shoot. The peduncle and inflorescence axis is prostrate just below the soil surface, with well-spaced flowers. Flower morphology is close to that of *Amomum*. Similarly, some species of *Amomum* (e.g., *A. biflorum*) can be confused with *Elettariopsis* in their vegetative habit.

Flowers of *Elettariopsis curtisii* appear solitarily, not in clusters, above the ground.

(Above) An inflorescence of *Elettariopsis triloba* has 5–10 flowers borne closely together. This species is widely distributed from Laos, Vietnam and Thailand to southern Burma and Peninsular Malaysia. (Left: Fruit.)

5. *Etlingera*
(including the genera *Nicolaia, Achasma* and *Geanthus,*
the last not found in Peninsular Malaysia)

As in *Alpinia*, a closer study of these gingers in their main distribution area in Indonesia has shown that the formerly established genera have to be merged as their distinguishing characters cannot be maintained. They are gingers with tall, leafy shoots and inflorescences on separate side shoots that may be long-peduncled or just appearing at soil level, or even partly covered in the ground, sometimes some distance from the leafy shoot.

Etlingera metriocheilos: red lips with white lower margins and black stigmas. (Inset: Fruit.)

Gingers of Peninsular Malaysia and Singapore

(Right and below) The incredible beauty of *Etlingera triorgyalis* inflorescences brighten up any forest foray: the immaculately arranged flowers have each a conspicuous white stigma, and the head of fruits, itself decorative from the ornamental ridges on the fruits (inset), is equally interesting.

Alpinieae: Etlingera

The best diagnostic characters are the inflorescence that is surrounded by a conspicuous involucre of sterile bracts, and in the floral structures, where the bracteoles are always tubular and the distal part of the lip becomes stiffly incurved after flowering; furthermore, the filament is very short, and the whole stamen much shorter than the lip. The flower colours range from pink, blood red to somewhat orange. The genus is still badly in need of revision, after more field studies. One species is widely cultivated, as a number of varieties: *Etlingera elatior*, the Porcelain Flower or Torch Ginger.

(Above) *Etlingera littoralis* flowers with narrow lips completely bounded by a yellow margin. The species is one of the earliest gingers named, described as early as 1792 by the Danish botanist Koenig from Phuket island (Thailand).

(Above) *Etlingera littoralis*: an inflorescence with completely red flowers.

(Above) The inflorescences of *Etlingera littoralis*, which may emerge through the ground a little away from the leafy shoots, give the impression of a bunch of flowers not connected to any plant.

(Opposite) *Etlingera punicea*, at the University of Malaya's Rimba Ilmu botanical garden.

Alpinieae: Etlingera

A spectacular display of *Etlingera venusta* inflorescences, each an alluring assemblage of flowers surrounded by showy, rose-pink bracts.

Like weapons of medieval warfare, these clusters of horn-like *Etlingera* fruits look additionally striking from their deep red colour.

Etlingera venusta, a close relative of the Torch Ginger (*E. elatior*), is found in Peninsular Malaysia and southern Thailand, where it usually flowers in April and May.

Alpinieae: Etlingera

(Above) *Etlingera maingayi*: fruiting head.

(Left and above) Flowering head of *Etlingera maingayi*.

(Opposite) *Etlingera maingayi*, widespread in Peninsular Malaysia and Singapore, has leafy shoots that are often raised above the ground on stilt roots. It has the smallest inflorescence heads (less than 4 cm across) among the peduncled species.

Gingers of Peninsular Malaysia and Singapore

(Above) *Etlingera hemisphaerica*, young fruits.

(Left) *Etlingera elatior*: a massive, knobbly head develops atop the robust scape as the fruits develop.

(Opposite) The Torch Ginger, *Etlingera elatior*. Cultivated throughout the tropics, it seems to be indigenous to Peninsular Malaysia. The many cultivated forms, varying in flower colour, are all extremely attractive: the large inflorescence (on a scape reaching 1.5 m long!) and showy, sterile bracts at the base of the flower-bearing part of the inflorescence make this species easily recognizable.

Etlingera elatior. (Painting by Charles de Alwis. Reproduced with kind permission of the Singapore Botanic Garden)

6. *Geocharis*

This is a small genus of seven species with two recorded for Peninsular Malaysia and another five reported for Sabah and Sarawak and the Philippines. It is believed to be closely allied to *Alpinia* and *Riedelia*, the latter restricted to New Guinea. Leafy shoots moderately tall with crossbars between the ribs of the leaf sheaths. Inflorescence separate from the leaves; rachis erect or curved, bracts small, flowers borne singly. Lip narrow, deeply bilobed almost to the base. Staminode rudimentary. Capsule fleshy, ridged or verrucose.

7. *Geostachys*

Forest herbs, 2–3 m high with a stout rhizome that is never below the ground but most often raised above forest floor on thick stilt roots. Most species are found in the Malay Peninsula, with a few in Indo-China, Sumatra and Borneo. The leaves in most species are narrowly elliptic. The inflorescence is on a separate side shoot, never buried in the soil but with short peduncles covered with stiff sheaths, rachis erect or bent slightly downwards. The flowering bracts are loosely overlapping and not like a cone-like structure. The flowers are yellowish to straw-coloured, often with

Geostachys cf. *megaphylla*, which grows up to 4 m high, is apparently restricted to the Cameron Highlands area in Peninsular Malaysia.

some red stripes or blotches; the labellum is 3-lobed at its tip, slightly longer than the corolla lobes. There are no lateral staminodes. Thirteen species are described from Peninsular Malaysia, three from southern Thailand. The species of this genus in Peninsular Malaysia are basically mountain plants. There may still be undescribed species.

Geostachys elegans.

(Right) *Geostachys taipingensis*, endemic to the Taiping Hills in Peninsular Malaysia.

(Above and left) *Geostachys decurvata*: strongly deflexed inflorescence near the end of flowering.

8. Hornstedtia

Medium-sized, or more commonly very tall gingers, some of the common species reaching a height of 7 m or more. The leafy shoot is coarse, often swollen at the base, with a diameter of up to 6 cm. The underground rhizome is coarse, in some species just below the surface, in others deep in the ground. The inflorescence, arising on a separate side shoot from the rhizome, is somewhat spindle-shaped on a very short peduncle. The involucral bracts are closely overlapping, stiff, often reticulate or ribbed, dark red. The flowers are red, in some species with cream or yellow margins, emerging a few at a time from the tip of the spindle-shaped inflorescence. There are no lateral staminodes, and the corolla lobes and the labellum are about the same length. The most common species are recognized by the spindle-shaped inflorescence with striate or reticulate nervation, a character that remains even in the dried state.

Hornstedtia scyphifera (Painting by Charles de Alwis. Reproduced with kind permission of the Singapore Botanic Garden)

(Opposite and overleaf) *Hornstedtia scyphifera* var. *grandis*.

9. Plagiostachys

Tall herbs easily recognizable by the inflorescence breaking through the leaf sheaths and thus appearing lateral on the leafy shoot, while in fact it is terminal. The dense inflorescence in some species has a few branches at its base (i.e., somewhat similar to the paniculate structure found in some *Alpinia*). It is a poorly known group of gingers due to the early disintegration of the inflorescence in many species into a mucilaginous mass that makes studies of herbarium specimens difficult or impossible. There are three species of *Plagiostachys* recorded for Peninsular Malaysia.

Plagiostachys albiflora.

8
Plant Chemistry in relation to Classification

Conventional studies of the Zingiberaceae based on morphology (plant form and structure) have been attempted by many botanists since the early days of Linnaeus. Following Linnaeus, others, such as Koenig, Roxburgh, Roscoe, Blume, Griffith, etc., have tried to unravel the complexities of the group. However, in as far as the gingers of Peninsular Malaysia are concerned, the work of Schumann, Ridley, Valeton and Holttum have contributed much to the understanding of the ginger flora. There is no doubt that Holttum's work (1950) is still considered one of the most comprehensive to date. Many of his descriptions of the floral parts, for instance, were very accurate.

However, all the revisions so far, including the more recent work of Smith (1985, 1986, 1987, 1988, 1989, 1990) have been based primarily on morphological characters. Not much has been studied about the chemical constituents of gingers in relation to their classification, i.e., chemotaxonomically. It is, however, known that such information provides useful tools that can aid conventional taxonomy. Isoenzyme analysis of selected species of the genera *Curcuma* and *Zingiber*, in particular of the peroxidase and esterase enzyme systems, have showed fairly distinct differences between species and slight variations within species (Ibrahim *et al.*, 1989; Ibrahim, 1996). For the true ginger (*Zingiber officinale*), it has also been established that the variation between the typical form and one of its races (locally known as *halia bara*, which has smaller red rhizomes) is hardly significant as shown by the peroxidase isoenzyme of the leaves.

It is interesting to note that species which are closely allied morphologically (at least in their overall vegetative characteristics), such as *Zingiber zerumbet* and *Z. ottensii*, differ only very slightly in their peroxidase isoenzymes of the rhizomes. Likewise, *Curcuma xanthorhiza* and *Curcuma zedoaria*, which are similar in their vegetative characters, are also almost identical in their esterase and peroxidase isoenzymes of the leaf extracts. In addition, malate and glutamate oxaloacetate transaminase (GOT) isoenzymes are found to be monomorphic in the *Curcuma* species studied. Several other enzyme systems such as isocitrate dehydrogenase, 6-phosphogluconate dehydrogenase and malate dehydrogenase may be monomorphic for some plants and not for others.

Members of the Zingiberaceae are usually aromatic in all or most parts or at least one of the plant parts. Many species are known to be rich in terpenoids; other compounds such as alkaloids and phenolics are not well documented. Phytochemical screening of selected species have revealed that flavonoids and terpenoids are ubiquitous but alkaloids were detected in three species of *Alpinia*, *Hedychium* and two species of *Zingiber* (Zakaria & Ibrahim, 1986). The presence of alkaloids in gingers is an interesting finding, as alkaloids are known to be more common in dicotyledons than in monocotyledons. Although saponins are characteristic of *Costus* spp. (Costaceae), Merh *et al.* (1986) found the presence of saponins in various genera of Zingiberaceae, such as *Globba*, *Curcuma*, *Hedychium*, *Zingiber*, and *Alpinia*, among others.

Volatile constituents have been used as taxonomic characters, especially at the generic or family level. Village folk and natives often recognise plants by the smell of crushed leaves, roots or other plant parts. In studying Zingiberaceae, with much field experience and knowledge one can sometimes identify the gingers at least up to the generic level from the smell of crushed leaves or rhizomes in non-flowering specimens.

Chemical studies have also revealed that gingers are rich in essential oils. Analysis of essential oils obtained from steam distillation of three species, namely *Zingiber spectabile*, *Z. officinale* (the red-rhizomed race) and *Alpinia galanga* showed a total of 34 compounds with 18 compounds, 20 compounds and 31 compounds identified for the three species, respectively. The most common compounds present in the three species studied included, among others, α-pinene, β-pinene, limonene, and β-elemene, etc. The three species also share many similar compounds, such as P-cymene, camphor, 1, 8-cineol, citral-a, linalool, β-

caryophyllene, α-humulene and β-bisabolene. The major compounds identified were trans-d-bergamotene from *Zingiber spectabile*, α-terpineol from *Alpinia galanga* and bornyl acetate from *Zingiber officinale* (Ibrahim & Zakaria, 1987). However, some variation in essential components of a species may occur due to its age, geographical location and climatic conditions.

In another study, two variants of *Elettariopsis triloba* were investigated for their volatile components (Mustafa *et al.*, 1996). Results showed that the two variants showed differences in their volatile constituents and variations were also detected when different plant parts (leaf, rhizome, root) were compared from the same plant. Some examples of the major components in the leaves of one variant of *Elettariopsis triloba* included α-citral, β-citral, 2,7-dimethyl-2, and geranyl isobutyrate; while its rhizome contained limonene and 2-carene. The percentage of volatile components in the roots was low. Similarly, examples of the major components of the leaves of a second variant of *E. triloba* included caryophyllene and eremophilene; while rhizomes contained cineole and borneol; the roots were found to have camphene and caryophyllene. Overall, it appears that a larger number of similar components were detected between roots and rhizomes as compared to the leaves of the two variants of this species.

There is still much research to be done, but further studies of the chemistry of gingers hold promise in clarifying the limits of some species and groups. This will be a useful tool that supports the classification of gingers, thus far confounded by the variation in certain groups, which presents difficulty in classifying them using conventional methods.

9

The Study and Collection of Gingers

Many taxonomists who have diligently worked on gingers will agree that these plants are better studied from living material than from dried specimens. The descriptions and measurements combined with good clear photographs acquired from live plants will be more accurate and useful to the field worker.

For many groups of Zingiberaceae our knowledge is still insufficient even about the basic morphological characters. This, in many cases, makes determination difficult. Much old herbarium material deposited in the large European herbaria in the Royal Botanic Gardens of Edinburgh and Kew, the National Herbarium in Paris and the Herbarium at the University of Leiden in The Netherlands, and also the two old herbaria in Singapore and in Bogor, Indonesia is rather difficult to use as flowers are often lacking or cannot be properly interpreted. Older collectors were drying their material over too long a period or using liquid preservatives for keeping collections in the field, both of which destroy the delicate flowers of this family.

Holttum (1950) and more recent taxonomists agreed that the structure of the inflorescences and flowers is of prime importance in the classification and identification of the Zingiberaceae. The structure of the inflorescence in *Globba* and *Alpinia* may not retain its perfect form in herbarium specimens. The bracts and bracteoles of some species of *Alpinia*, for instance, are not persistent or long-lasting and will usually drop before the onset of flowering. In such cases the accuracy or the completeness of herbarium specimens will therefore depend on the time of collection. Hence careful notes on the bracts or bracteoles must be included in the field notes. Sometimes, minute features of the flower or floral parts may have been damaged or not easily detected in dried or preserved specimens. These features can be of taxonomic significance. For instance—the *lip* or *labellum* (colour, especially of the mid-lobe, spots, bands, shape, size); *anther appendage* (colour, number, the latter important in *Globba* in particular); *lateral staminodes* (colour, and if reduced to rudimentary structures and not easily detected); *corolla lobes*, *calyx* and *bracts* (colour, shape); *leaf* (colour, variegations, purplish flush on the lower leaf surface, smell of crushed tissues); *petiole*, *ligule* and *sheath* (colour, pubescence); *rhizome* and *roots* (colour, smell); and *fruits* (colour, smell).

Modern specialists in the Zingiberaceae first select their material carefully, then make colour photographs of the plant with 'close up' of the flowers and of course at the same time make extensive notes about underground parts (rhizomes and roots), height of the shoots, search for fruits and finally collect some individual flowers in alcohol for further studies.

After anthesis the dense inflorescence of *Plagiostachys* is more or less turned into a mucilaginous mass that obscures the structure. Herbarium material of inflorescences is therefore most often useless even for naming a sample. Similar difficulties face the specialist in *Curcuma* and *Zingiber*. In *Curcuma* the inexperienced collector often mistakes the coma (the large coloured, sterile bracts terminating the inflorescence in most species) for flowers and collects plants that have long passed the flowering stage.

In fact every genus imposes special requirements for the collector, e.g., in the genus *Globba* it is very important to have careful collections of the anthers as one of the very important characters is the number and position of the lateral anther appendages. In other genera, such as *Kaempferia*, the apical anther appendage (the anther crest) is an important character.

In cone-like inflorescences such as in *Zingiber*, *Hornstedtia* or bulky inflorescences such as in some species of *Amomum* and *Plagiostachys*, the inflorescence should be cut in half before being pressed. Ideally a duplicate of the inflorescence should also be preserved in alcohol or spirit with the individual flowers preserved separately in smaller tubes.

The plane of distichy of leaves on the shoot is a reliable feature in distinguishing the tribes. The plane of distichy of leaves is parallel to the rhizome in the tribes Hedychieae, Globbeae and Zingibereae and transverse to the rhizome in the Alpinieae. This, however, is not always easy to determine.

Material for herbarium material is still needed, and if the population is large enough it is advisable to collect duplicates to distribute among the centres for research in the family. If the shoot is large, e.g., 3–5 m, as is the case of many rain forest species, typical leaves from the middle part of the pseudostem is collected. Preparation of herbarium specimens for smaller gingers such as species of *Kaempferia*, *Camptandra*, *Globba*, etc., should be easy as the whole plant can be carefully pressed to dry.

It is also important to make notes about the base of the shoot, which may be swollen or have specially developed leaves or leaf-sheaths. Colours of leaves (upper and lower side) are also noted, as these characters often change in the drying process. If the specialist has access to a botanical garden willing to establish a living collection of gingers, then rhizomes should be collected for growing and further studies, e.g., on cytology, palynology and molecular studies.

The ideal approach in studying gingers is perhaps to be able to bring the rhizome back for cultivation and observation. This can yield interesting information. For instance, in many species of *Zingiber* the height of the peduncle and the colour of the inflorescence bracts change at different stages of the inflorescence. Such information is rarely available on notes accompanying herbarium specimens.

On the contrary, the study of garden-grown plants is not always possible as some species from the genera such as *Hornstedtia*, *Geostachys* and other high-elevation gingers will not survive in lowland climatic or soil conditions.

10
Notes from a Ginger naturalist

Ginger-hunting in the rain forest: 'Belum' Gingers

In the botanically rich rain forest areas of Peninsular Malaysia, "plant hunting" can be rewarding. One of the more recently discovered excitements is the Belum rain forest, more accurately, the Temengor Forest Reserve in the state of Perak in Peninsular Malaysia. The Belum rain forest covers about 290,000 hectares at the northern tip of Perak, just south of Thailand. The Belum forest is rich in gingers and botanical forays there reveal the variety of plant form and habit that can be encountered.

Members of the Zingiberaceae form an important component of the herbaceous flora in Belum. Several visits there between 1994 and 1996 have yielded 34 species in 10 genera. These represent some 23% of the 150 species recorded in Holttum's 1950 account and about 56% of the 18 genera of Zingiberaceae in Peninsular Malaysia. More work needs to be done on the gingers in other areas of the vast forests of Belum and not all those recorded so far have been named or fully identified.

The area around the Base Camp (230 m elevation) at Sungei Halong up to Bukit Kabut (1300 m) is particularly rich in gingers, with almost two thirds of the total encountered found at this site.

Most of the ginger species grow in the humid, shady undergrowth, their leafy shoots emerging among the wet leaf litter of the forest trees. Some species grow by the banks of shaded small streams while others, being more light-tolerant, grow by the banks of more open or unshaded streams. A special delight is provided by *Etlingera triorgyalis*, which produces spectacular scarlet flowers at the base of its leaf shoot, giving a brilliant display among the dull, dark brown leaf litter in the gloom of the forest floor. Still others grow on steep slopes or are found scrambling over rocks, their roots dipping into small pockets of soil in crevices,

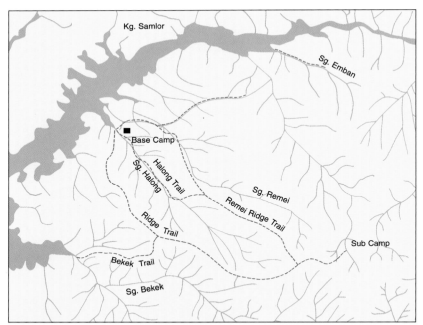

The area in the Belum rain forest, Peninsular Malaysia, that was specially visited for gingers.

finding support. *Hedychium longicornutum* with long, curly, yellow to orange scarlet flowers, is the only epiphytic ginger in Belum. It attaches itself to its supporting tree by long, succulent, clasping roots, not far above the ground.

The majority of Belum gingers are herbs 2–4 m tall. However, *Hornstedtia ophiuchus*, with hard, somewhat "woody" stems is exceptionally tall, measuring over 7 m although the average height of the species is 4 m. This was found growing along the trail from Sungei Halong to Bukit Kabut. Among the smaller gingers, measuring 1 m or less, are *Globba cernua* and *Elettariopsis smithiae*, the latter in both its typical form and as the variety *rugosa*. Even shorter and smaller, and without a pseudostem, the leafy shoots thus appearing as tufts, is *E. curtisii*. Some gingers grow as solitary shoots while others form clumps. Some others, like *Scaphochlamys kunstleri*, grow in low tufts, forming carpets among the herbaceous ground flora in the forest.

Globba. *Globba* species are plants of shady forests. They are small, less than 1 m tall. They may grow singly or in a group; sometimes the leafy shoots are formed so closely that they superficially resemble a rosette. Globbas frequently produce bulbils or vegetative miniature ginger plantlets at the lower end of the inflorescence. The new plants which grow up from these bulbils will have the same features unchanged from those of their parents. In Belum this proves to be an efficient way of dispersal as *Globba cernua* is distributed all along the Sungei Halong–Bukit Kabut trail. In some localities the bulbils could have been washed away to establish elsewhere during heavy rainfall. Fruits when formed, though infrequent, are globbose and rugose (warty). So far only this species has been encountered in Belum. It produces small flowers with a pale yellow lip at the centre of which is a bilobed olive brown spot. At the tip of the long filament is the anther with four pale brown triangular appendages.

Zingiber. *Zingiber* species can be divided into two types: one group produces flowers with a purple or pink lip mottled cream or pale yellow; the other group produces flowers with lips entirely white or cream, not mottled at all. Both groups are represented in Belum. A total of five species have been recorded so far, only two of which are identified.

Z. spectabile occurs here, but can be found throughout the Peninsula. The leafy shoot is usually about 2–3 m tall. The species has the largest ginger inflorescence of which the scape (inflorecence stalk) is up to 50 cm while the inflorescence (flower spike) is up to 30 cm. The yellow bracts, which turn red when old, form pouches containing mucilage in which the flowers develop. The lip of the flower is purple, mottled with pale yellow. In Belum they grow in several localities. However at one locality in Sungei Bekek, very tall plants more than 4 m tall, grow in large clumps, producing exceptionally large inflorescences with scapes up to 58 cm long and the inflorescence proper up to 44 cm long.

Another species, with a purple lip mottled cream, is yet to be identified. The leafy shoot of this *Zingiber* is about 2 m tall. It produces very long, slender, cylindric yellow inflorescences measuring 53–89 cm tall and 1.5–2 cm wide. The closely overlapping yellow bracts which turn pink red when old, do not form pouches. Both the mid and side lobes of the lip are deep purple, appearing almost black, mottled creamy white. The

Zingiber spectabile.

anther is white, while the anther appendages is deep purple. The colouration of the inflorescence and the flower resembles *Z. curtisii*.

Z. puberulum is from the group that produces entirely creamy white flowers. It is a tall *Zingiber*, the leafy shoots grow up to 3 m. The young inflorescence is prostrate, producing white flowers which emerge between the closely overlapping, red bracts. When flowering ends, the base of the inflorescence bends upwards so that the whole flower spike is upright as the fruits develop. This species is found around the Base Camp at Sungei Halong.

Two other unidentified *Zingiber* have been encountered in Belum. One resembles *Z. gracile* in leaf shape. The leafy shoot is about 2–2.5 m. In some of the plants, the tip of the leaf shoot, where only very small leaves are produced, bends over and as it touches the ground, produces young plants vegetatively. Sometimes bulbils are produced at the leaf axils near the tip of the leafy shoot, which grow to become young plants vegetatively. The inflorescence is ovoid, with red bracts. Another unidentified

Zingiber produces ovoid inflorescences with green bracts, turning pink when old, with entirely white flowers. It resembles *Z. zerumbet* but the leafy shoot is very hairy. The *orang asli* (indigenous people) use this species in their traditional medicine.

Curcuma. A single species, *Curcuma xanthorhiza*, found only at one location, near an abandoned *orang asli* settlement at Sungei Bekek, may not be a wild ginger but introduced into the jungle as it is known that the nomadic people of these forests visit Grik town for their groceries and may be in contact with villagers living nearby. This species was located in an open area, growing in a large clump. It is the largest species of *Curcuma* in the Peninsula, up to 2 m tall. Known locally as *temu lawak*, it is used in traditional medicine. The plant produces large leaves with a faint purple "feather" stripe on either side of the green midrib. The inflorescence is produced separately from the leafy shoot. The foliage leaves appear at the end of flowering. The flowering spike with its stalk, 22–45 cm tall, is rather colourful. The bracts at the base of the spike are pale green while those at the apex, forming the coma, are purple, both types forming pouches in which 1–6 yellow flowers appear. Each flower has a yellow lip with a darker yellow median stripe and is surrounded by three pink petals. In the throat of the yellow flower is the single stamen with two anther spurs.

Hedychium. Only one species, the epiphytic *Hedychium longicornutum*, has been found in Belum. It is common near the Base Camp at Sungei Halong. Their leafy shoots (1.5–2 m long) grow close together, forming a clump attached to a tree trunk, (much like an epiphytic fern or orchid) with long, thick succulent roots exposed like a writhing mass of white, fat worms. Indeed the *orang asli* use them in their traditional medicine as a vermifuge to deworm children. The base of the leaf shoot is bulbous, covered by red-purple leaf sheaths. The leaves are thick and leathery.

When the whole clump is in bloom it is spectacular. The inflorescence at the tip of each leaf shoot is a globose mass of entangled, curly, yellow and orange-red long petals and staminodes with their pale yellow to orange, long filaments and anthers radiating from it. In fruit, it is just as beautiful. The ripe, orange fruit burst open to reveal seeds covered by bright red, thread-like arils, contrasting with the orange inner fruit wall.

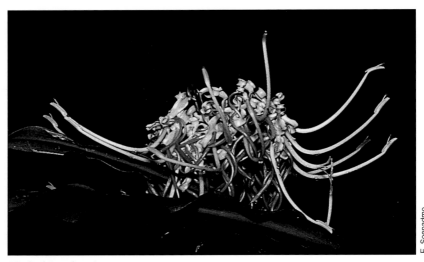

Hedychium longicornutum.

Scaphochlamys. Two species have been recorded in Belum. *Scaphochlamys kunstleri* is common in some localities, such as Bukit Kabut where it forms large patches. It also occurs at Sungei Emban and Sungei Bekek. The plant spreads with its fleshy, underground rhizome. The inflorescences with its pouches, especially in large plants, superficially resembles *Curcuma* and may be mistaken for it. The white flowers resemble an orchid, the lip is white, with a yellow median band and pink or crimson streaks on either side towards the base.

Another *Scaphochlamys* species has been located by a stream bank near the Sub-Camp at Bukit Kabut at 1000 m. The leaves are dark green with a purple lower surface. The white flowers, with a pale yellow median band, superficially resemble an orchid.

Scaphochlamys kunstleri.

Alpinia. Two species of *Alpinia* grow in Belum. The leaf shoots of both are 2–3 m tall. Similarly the two have large, prominently ribbed leaves and produce green, spherical fruits, measuring 2–2.5 cm in diameter. *A. javanica* was found growing in the undergrowth by the side of an old logging road. The erect inflorescence and its stalk, 29 cm tall, is striking with long, lateral branches (6 cm long), each drooping with 1–2 heavy, round fruits crowned by its brown, funnel shaped, persistent calyx. The inflorescence of the other *Alpinia* sp. is without such long lateral branches.

Etlingera. Of all the ginger species in the Peninsula, those of the genus *Etlingera* produce the most colourful inflorescences, flowers and fruits. They are represented by nine species in Belum. The inflorescences of the 'achasma' group of etlingeras emerge from the leaf litter to have the long lips of their flowers collectively appearing as a brilliantly coloured single flower with radiating petals on the forest floor. Holttum (1950) recorded six species in the Peninsula, with lips which are either entirely red, red with a yellow median band or else red with yellow or white margins. In

Etlingera littoralis.

Belum five of these have been located, *Etlingera punicea*, *E. metriocheilos*, *E. triorgyalis*, *E. littoralis* and *E. pauciflora*. The last, as its name suggests, has few flowers. Only 1–3 are formed in each inflorescence. The fruit heads of the first three etlingeras mentioned above are spherical, large, about 8 cm in diameter, with persistent bracts at their bases. Each individual pink or red fruit is ridged with the scar of its calyx at the apex. *E. pauciflora* has small, smooth, creamy white fruits which are not ridged. Its whole fruit head measures about 3.5 cm depending on the number of individual fruits formed. The fruits of *E. metriocheilos* are not smooth, as described in an earlier account, but instead are ridged.

The 'nicolaia' group of etlingeras produce their inflorescences at the ends of the tall leafless shoots and not at ground level as do the achasmas. Four species have been recorded for the Peninsula. They are *Etlingera maingayi*, *E. hemisphaerica*, *E. elatior* and *E. venusta*. *E. elatior* is the familiar kantan used locally for flavouring food. It has not been located in Belum. However, four etlingeras of the "nicolaia" group have been found growing in Belum, so that one of them is either a new record or species. Its identity is yet to be determined. Vegetatively it resembles *E. venusta* except that its leaves are always sessile while those of *E. venusta* have a distinct petiole of up to 3 cm long. In the shape of the deep involucral cup, this unidentified species also resembles *E. venusta* but in the colour of the involucral and floral bracts, it is pink like those of *E. elatior*. There is also a difference in the colour and shape of the floral bract which is pink and acute at the apex while that of *E. venusta* is red and bluntly rounded. The lip of the flower of both species is elliptic and pale pink with a red median patch. As the large fruit head (composing of more than 13 fruits) matures, all the involucral and floral bracts decay while the fruits of *E. venusta* mature in their persistent involucral cups. The flowers of both species are smooth and red with long beaks.

E. hemisphaerica is recorded by Holttum as being known from few collections. Its broad, shortly spreading, red involucral bracts forming a shallow cup, are distinctive. The flowers are rose red with a yellow margin. At maturity the large, light green fruit head is globose, slightly flattened at the top. Each fruit is short hairy with a long black beak, the remnants of the calyx. Vegetatively the leafy shoot is most attractive. It is slightly smaller than *E. elatior*. The leaves, with a deep purple lower surface which fades at maturity, are glossy dark green at the upper surface and with a wavy leaf margin.

E. maingayi has the smallest inflorescence of the five species, forming a compact head of deep pink flowers with red centres. The bracts of the inflorescence appear white or pale pink due to their hairy surfaces. The fruit head at the end of a long, slender, red stalk is a compact head of smooth, shiny, scarlet fruits. Each fruit is topped by its persistent calyx. The whitish pink bracts remain at the base of the mature fruit head.

Amomum. *Amomum* is represented by at least five species in Belum. *A. lappaceum* grows near the Base Camp at Sungei Halong and is common at Sungei Bekek and Sungei Emban. As in a number of *Amomum* species, its inflorescence gradually elongates as it continues to flower at the apex while the fruits develop at the base. The weird-looking, brown club-shaped inflorescence covered with mucilage, emerges from the leaf litter, contrasting with the yellow flowers at the apex. The fruits are ovoid, green, covered by soft, fleshy spines. The elongated bunch of mature fruits may finally be more than 30 cm long, looking like a narrow bunch of small green *rambutan* fruits.

A. testaceum is another species with a long inflorescence stalk. The cylindric inflorescence is covered by light brown, papery bracts. The flowers are white or cream with a patch of yellow at the centre. The fruits, formed among the persistent papery bracts are small, about 1.5 cm in diameter, round, with brown seeds embedded in a white, sweet pulp. The *orang asli* collect and trade the fruits among themselves.

Amomum lappaceum.

Amomum uliginosum.

The other three *Amomum* species produce flower heads partially embedded in the forest floor. *A. uliginosum* has a large, white, concave lip with a yellow median band and a dark crimson line on either side. The mature fruit head is spherical, composed of small, round fruits about 1.5 cm in diameter, its surface covered by soft, red spines. The whole bunch of fruits, or fruit head, looks like a small bunch of red, spiny rambutan.

Of the other two *Amomum* species, *A. aculeatum* has yellow flowers with crimson spots and streaks. The anther has at its apex a crescent-shaped crest, the upper margin of which is three-lobed. The leafy shoot, about 2.5 m tall, has glossy dark green leaves.

The fifth *Amomum* is *A. hastilabium*. Its flower has a semi-circular white lip, with an orange median band with crimson streaks on either side. The fruits are hairy, round, ridged, pale buff, formed among light brown papery persistent bracts.

Hornstedtia. Of the two *Hornstedtia* species growing in Belum, *H. ophiuchus* mentioned earlier, is exceptionally tall (measuring more than 7 m tall) and found along the trail from Sungei Halong to Bukit Kabut.

It also grows at Sungei Bekek. It has spindle-shaped inflorescences. The bracts, tightly overlapping and tough, are covered by white short hairs except for a broad red margin towards the apex. It has red flowers which emerge from the tip of the spindle. Holttum (1950) stated that *H. ophiuchus* is only known from the original collection of Ridley made in July 1891, from the Tahan River (Pahang state). *H. conica* is much smaller, 2–3 m tall. The leafy shoots growing at the Ridge trail near the Base Camp produce glaucous leaves (they are covered by a blue-green bloom) with red margins and are most attractive. The flower spikes are spindle shaped, narrowed to a point, from which one or two flowers emerge. The bracts of the flower spike are covered by white hairs except towards the dull red to purple edges. Each flower has an elongated lip, crisped at the edges, pink with two red lines at the centre. As the fruits develop, the bases of the spindle shaped inflorescence broaden, becoming characteristically onion-shaped.

Elettariopsis. This is a small genus of four or five species recently revised and described by Kam (1982). They are small ginger plants, not more than 1 m tall. Three species grow in Belum, two, *Elettariopsis curtisii* and *E. triloba* release strange, stink-bug odour when their leaves are bruised. The *orang asli* use both in their traditional medicine. Both species hold their elliptic leaves erect. The leafy shoot has 1–5 leaves. The former has a short leaf tip, the latter has a leaf tip of up to 3 cm long. Both produce small white flowers, the lip with a yellow median band and a red stripe on either side. The flowers of *E. triloba* are produced in a compact head at the base of the leaf shoot, while those of *E. curtisii* are formed on a long horizontal, branched inflorescence. Both species have been located near the Base Camp at Sungei Halong. A third species *E. smithiae* is more widely distributed, and has been recorded from Sungei Halong, Sungei Emban, Sungei Bekek and other sites. The leafy shoot has a distinct pseudostem with 3–8, sometimes 9, grey green, floppy leaves more or less arranged in two ranks. The flowers, with the same white, yellow and red pattern as the previous two species above, are produced on a long, branching inflorescence. The fruit is a pale brown, globose capsule, slightly ridged, about 3 cm in diameter, formed among the leaf litter. At the Ridge Trail and at the entrance to the Base camp, is a variety named *E. smithiae* var. *rugosa* which differs from the typical form in having broader leaves whose surfaces are conspicuously crinkled.

11
A Ginger Garden

Members of the ginger family, Zingiberaceae, can be successfully grown as garden ornamental plants. Their interesting forms add variety to the garden scenery; besides, their leaves have fragrant tissues and some species have perfumed flowers.

The tall leafy species can be grown at the edges of the garden as a screen in place of trees or beneath them as they do in their natural environment in the forest. Most of the tall species seem to be tolerant of the hot sun. Examples of these are the 'achasma' and 'nicolaia' *Etlingera* species, with their spectacularly gorgeous flowers at the bases of their leaf shoots. Well grown *Zingiber spectabile*, with long, cylindric yellow or red inflorescences, can also make a decorative screen or fence.

Some other ginger species are smaller, from less than 1 m to 1–2 m tall. Their stems often grow close together forming clumps with gracefully spreading leaf shoots. While some of these smaller gingers can be grown in the open, e.g., *Zingiber kunstleri* and *Z. ottensii*, most do better in light shade. *Alpinia vitellina* with orange yellow flowers at the ends of their leaf shoots and *Zingiber gracile* and *Z. griffithii* with yellow, pink or red cylindric flower spikes at the bases of their leafy shoots are examples.

The globbas, all small plants, less than 1 m tall with small, dainty flowers, grow best in the shade or else their leaves curl up. Requiring similar conditions of shade are the round, large-leaved *Scaphochlamys* and *Boesenbergia* which form 'carpets' on the forest floor. These species can be grown beneath the tall ginger plants.

The smallest species are the low, ground-hugging plants like the familiar *Kaempferia galanga* (*cekur*), used for flavouring food and *K. pulchra*. These can be used as ground covers since they spread rapidly by their underground storage stems and their leaves lie flat on the ground. The latter has a most attractive brown and green leaf pattern and lilac flowers.

The epiphytic ginger, *Hedychium longicornutum*, with colourful, showy inflorescences, described elsewhere in this book, makes an interesting addition to the garden. It has specialised roots adapted for attaching the plants to a support. This ginger can be easily grown by tying the base of the leafy shoot to a tree trunk. The upper part of the leafy shoot, which may be 1–2 m tall, also needs to be supported. The plant has to be watered frequently for new roots to form.

Hedychium longicornutum can also be grown easily by placing a leafy shoot at the fork of a branch and covering the base of the leafy shoot with decayed leaf mould or leaf litter. Watering it liberally and frequently will soon produce new roots and shoots. A third method is by placing the leafy shoot in the growing centre of a nest-fern, *Asplenium nidus* and similarly covering the base with a mulch of dead leaves. Again frequent watering produces new roots which grow into the huge mass of fibrous roots of the nest-fern, which does not seem to be harmed. This last method arose from an observation by Christopher Wells who saw *H. longicornutum* flowering from a nest-fern, epiphytic on an *Araucaria* tree in Fraser Hill.

All ginger plants grow best in wet humid conditions. To grow them successfully out of their jungle environment, spraying often is necessary although inconvenient. The penalty for forgetting to spray them, especially the tall species, on a hot day, will be the distressing sight of these graceful shoots bending over and destroyed. They will not recover and have to be cut. The reason is that the true shoot is actually made up of a few thin-walled stem cells, surrounded entirely by overlapping leaf sheaths without any supportive, woody tissues, forming the pseudostem. Many of the medium to tall species grow beside small, forest streams with their much branched, fibrous roots in the clear, flowing waters without any apparent harm. In the garden, however, it may be better to grow them in well-drained soil.

A convenient way to achieve this is to place the plant at soil level, without digging a planting hole and covering the rhizome and roots with a mixture of loam or garden soil, sand and compost. The latter is used as a soil conditioner, moisture retainer and as a slow-release fertilizer. This method of planting was first suggested by A. Santiago. Most ginger species can be grown by this method, directly on the surface of the ground, including tall types like the etlingeras and the hornstedtias. Frequent watering of this porous mixture, at the same time water-retaining, is necessary and new roots form quickly.

Providing this wet environment is extremely critical if ginger plants just taken out of the jungle are to be successfully grown out of their natural habitat. The rhizomes of most ginger plants are not perennating or resting organs. They do not contain large amounts of stored food. Therefore rhizomes, collected from jungle plants, must be quickly planted and new shoots and roots encouraged to grow, new green leaves to be formed, able to photosynthesise, before their stored food supply is exhausted. In fact, new shoots, grown from cut pieces of rhizomes without leaf shoots attached to them to provide food continuously, are disappointingly puny at first. However, if these new, feeble shoots are then provided with a fertilizer and can establish themselves, subsequent shoots will be progressively larger and stronger. In about a year, tall plants like the wild ones will be established.

In contrast to the non-perennating rhizomes of most gingers are the fleshy, resting organs of the domesticated ginger *Zingiber officinale* and all the *Curcuma* species, including *C. domestica* (*kunyit*) which contains large amounts of stored food. New plants formed from these resting rhizomes are robust. To grow these, a special bedding 10–30 cm high may be prepared, using the same mixture of garden soil, sand and compost as already described above. For small gingers like *C. domestica* and *Z. officinale*, a small bed of soil 10 cm high may be enough to cover the rhizomes completely. The former seems to flower irregularly locally while the latter rarely or not at all.

Bigger species like *C. zedoaria* (*temu kuning*) and *C. mangga* (*temu pauh*) need a deeper bed of 15–20 cm. The largest curcuma in the Peninsula, *C. xanthorhiza* (*temu lawak*) requires a mound or bed of 25–30 cm high. With frequent watering, the compost quickly deteriorates and has to be replaced 3–4 times a year. In order to encourage flowering, the rhizomes should be left to rest after the growing season. All the aerial shoots are to be cut off when the leaves turn yellow. The rhizomes should be left to rest in the soil undisturbed for about 3–4 months. Watering is reduced to a minimum, just enough to keep the soil moist but not sufficient for aerial, vegetative shoots to grow. After resting, with copious water provided for growth, *C. xanthorhiza*, *C. zedoaria*, *C. aeruginosa* and *C. mangga* produce flower spikes on leafless shoots. Leaf shoots grow towards the end of flowering at the bases of the flower spikes. For *C. domestica* and *C. aurantiaca* the inflorescences are formed in the middle, i.e., the apex of the leafy shoot.

Bulbils or miniature ginger plantlets are frequently produced by the globbas at the bases of their inflorescences. These can be picked off and planted just like any rhizome. Similarly some zingibers may also produce bulbils at their leaf axils, especially those at the top of the leaf shoots.

New plants can also be grown from seeds of most ginger species. The seeds can be collected after flowering and waiting for the seeds to mature, which can take up to two months in the bigger ginger species like the achasmas. They are just planted out like seeds of ordinary plants in a seed bed. Species of *Curcuma*, *Kaempferia*, *Zingiber* and numerous other species can be propagated this way.

The growing medium of ginger plants must not be allowed to dry out. To keep the soil cool and moist, a mulch of dead leaves is placed around the bases of the leaf shoots. A ready supply of leaf litter may be conveniently obtained by growing ginger plants beneath leafy trees like the cinnamon, *Cinnamomum iners* (*kayu manis*), commonly planted as a road-side tree. If the leaf litter is insufficient, lower branches of the tree may be cut and left to dry. The leaves are then stripped and used as a mulch for the ginger plants. Similarly the dead leaves of other trees in the garden like *nangka* (*Artocarpus heterophyllus*), avocado (*Persea americana*), etc., can be used. With the heat and frequent hosing, the mulch quickly disintegrates and has to be replaced every three months or so. Only dead leaves are used and not fresh grass cuttings as the latter generate too much heat which will kill the plants that they are supposed to cool.

Finally, a word about insect pests of gingers. Fortunately most gingers seem to be free of them. Occasionally leaf-roller caterpillars of small moths may eat up the leaves while hiding, rolled up in a tube of the leaf blade, much like the banana skipper. The leaves of globbas and *Hedychium coronarium* seem to be susceptible. Leaf-miners, tiny larvae which tunnel their way as they eat the tissues between the upper and lower surfaces of the blades, do not seem to do much harm, if only a few of them attack. More unsightly is the work of thrips, minute black insects, which deform and stunt leaf and flower growth by their feeding activity. Leaf margins which curl up or dry up usually betray their presence. Grasshoppers (*Valanga nigricornis*) chew up and eat quite a lot of leaves when several adults are present on a single plant. Sucking stink-bugs by their feeding habit may introduce viruses and bacteria during feeding. These insects emit a strong odour when disturbed.

The most serious pest seems to be the white mealy bugs and scale insects which suck plant juices. Young shoots wither and leaves become deformed and unsightly during their attack. These insects are usually brought by ants which tend to them (like people tending cattle) since they produce sugary secretions as they feed on the leaves. The ants can be seen feeding on the secretions. The sugary liquids also encourage sooty moulds to form black "mats" on the leaf surfaces.

To discourage all the above mentioned insects from causing damage to leaves or inflorescences, a general common and safe insecticide like malathion can be sprayed occasionally or only when the pests are abundant. Sooty moulds can be rid by spraying with a general fungicide.

Peninsular Malaysian and Singapore Gingers A Checklist

* not native

Alpinia
A. aquatica (Retz.) Rosc.
A. assimilis Ridl.
A. capitellata Jack
A. conchigera Griff.
A. corneri (Holtt.) R.M. Smith
A. denticulata (Ridl.) Holtt.
A. galanga (L.) Sw. *
A. javanica Bl. var. *colorata* Ridl.
A. javanica Bl. var. *javanica*
A. latilabris Ridl.
A. macrostephana (Bak.) Ridl.
A. malaccensis (Burm.) Rosc. var. *malaccensis*
A. malaccensis (Burm.) Rosc. var. *nobilis* (Ridl.) I.M. Turner
A. mollissima Ridl.
A. murdochii Ridl.
A. mutica Roxb.
A. oxymitra K. Schum.
A. pahangensis Ridl.
A. petiolata Bak.,
A. pulcherrima Ridl.
A. purpurata (Vieill.) K. Schum. *
A. rafflesiana Wall. ex. Bak. var. *hirtior* (Rid.) Holtt.
A. rafflesiana Wall. ex. Bak. var. *rafflesiana*
A. scabra (Bl.) Bak.
*A. seimundi*i Ridl.
A. vittelina (Lindl.) Ridl. var. *cannaefolia* (Ridl.) I.M. Turner
A. vittelina (Lindl.) Ridl. var. *vittelina*
A. zerumbet (Pers.) B.L. Burtt & R.M. Smith *

Amomum
A. aculeatum Roxb.
A. biflorum Jack.
A. cephalotes Ridl.
A. citrinum (Ridl.) Holtt.
A. cylindraceum Ridl.
A. hastilabium Ridl.
A. kepulaga Sprague & Burk. *
A. lappaceum Ridl.
A. macrodous Scort.
A. macroglossum K. Schum.
A. micranthum Ridl.
A. ochreum Ridl.
A. rivale Ridl.
A. spiceum Ridl.
A. squarrosum Ridl.
A. testaceum Ridl.
A. uliginosum Koenig
A. utriculosum (Ridl.) Holtt.
A. xanthophlebium Bak.

Boesenbergia
B. clivalis (Ridl.) Schltr.
B. curtisii (Bak.) Schltr.
B. flava Holtt.
B. longipes (King & Prain) Schltr.
B. plicata (Ridl.) Holtt.
B. prainiana (King *ex* Bak.) Schltr.
B. pulcherrima (Wall.) Kuntze
B. rotunda (L.) Mansf. *

Camptandra
C. latifolia Ridl.
C. ovata Ridl.
C. parvula (King *ex* Bak.) Ridl.

Curcuma
C. aeruginosa Roxb. *
C. aurantiaca van Zijp

C. colorata Val. *
C. domestica Val. *
C. mangga Val. & van Zijp *
C. parviflora Wall.
C. viridiflora Roxb.
C. xanthorhiza Roxb. *
C. zedoaria (Berg.) Rosc. *

Elettaria
E. cardamomum (L.) Maton *
E. longituba (Ridl.) Holtt.

Elettariopsis
E. burttiana Y.K. Kam
E. curtisii Bak.
E. exserta (Scort.) Bak.
E. smithiae Y.K. Kam var. *rugosa* Y.K. Kam
E. smithiae Y.K. Kam var. *smithiae*
E. triloba (Gagnep.) Loesen.

Etlingera
E. elatior (Jack) R.M. Smith
E. hemisphaerica (Bl.) R.M. Smith
E. littoralis (Koenig) Giseke
E. maingayi (Bak.) R.M. Smith var. *longibracteata* (Holtt.) I.M. Turner
E. maingayi (Bak.) R.M. Smith var. *maingayi*
E. metriocheilos (Griff.) R.M. Smith var. *grandiflora* (Holtt.) I.M. Turner
E. metriocheilos (Griff.) R.M. Smith var. *major* (Holtt.) I.M. Turner
E. metriocheilos (Griff.) R.M. Smith var. *metriocheilos*
E. metriocheilos (Griff.) R.M. Smith var. *petiolata* (Holtt.) I.M. Turner
E. metriocheilos (Griff.) R.M. Smith var. *rubrostriata* (Holtt.) I.M. Turner
E. pauciflora (Ridl.) R.M. Smith
E. punicea (Roxb.) R.M. Smith
E. subterranea (Holtt.) R.M. Smith
E. triorgyalis (Bak.) R.M. Smith
E. venusta (Ridl.) R.M. Smith

Geocharis
G. aurantiaca Ridl.
G. secundiflora (Ridl.) Holtt.

Geostachys
G. decurvata (Bak.) Ridl.
G. densiflora Ridl.
G. elegans Ridl.
G. leucantha B.C. Stone
G. megaphylla Holtt.
G. montana (Ridl.) Holtt.
G. penangensis Ridl.
G. primulina Ridl.
G. rupestris Ridl.
G. secunda (Bak.) Ridl.
G. sericea (Ridl.) Holtt.
G. tahanensis Holtt.
G. taipingensis Holtt.

Globba
G. albiflora Ridl. var. *albiflora*
G. albiflora Ridl. var. *aurea* Holtt.
G. cernua Bak. ssp. *cernua*
G. cernua Bak. ssp. *crocea* S.N. Lim
G. cernua Bak. ssp. *porphyria* S.N. Lim
G. corneri A. Weber
G. curtisii Holtt.
G. fasciata Ridl.
G. fragilis S.N. Lim
G. holttumii S.N. Lim ssp. *aurea* S.N. Lim
G. holttumii S.N. Lim ssp. *holttumii*
Globba × *intermedia* S.N. Lim (endemic hybrid)
G. leucantha Miq. var. *bicolor* Holtt.
G. leucantha Miq. var. *flavidula* (Ridl.) Holtt.
G. leucantha Miq. var. *peninsularis* Holtt.
G. leucantha Miq.var. *violacea* (Ridl.) Holtt.
G. marantina L.
G. nawawii H. Ibrahim & K. Larsen
G. patens Miq. var. *costulata* S.N. Lim

G. patens Miq. var. *patens*
G. pendula Roxb. ssp. *montana* (Ridl.) S.N. Lim
G. pendula Roxb. ssp. *pendula* var. *elegans* (Ridl.) Holtt.
G. pendula Roxb. ssp. *pendula* var. *pendula*
G. schomburgkii Hook.*
G. unifolia Ridl.
G. variabilis Ridl. ssp. *pusilla* S.N. Lim.
G. variabilis Ridl. ssp. *variabilis*

Haniffia
H. cyanescens (Ridl.) Holtt.

Hedychium
H. chrysoleucum Hook.
H. collinum Ridl.
H. coronarium Koenig *
H. hirsutissimum Holtt.
H. longicornutum Bak.
H. macrorhizum Ridl.
H. malayanum Ridl.
H. paludosum Hend.

Hornstedtia
H. conica Ridl.
H. leonurus (Koenig) Retz.
H. ophiuchus (Ridl.) Ridl.
H. phaeochoana (K. Schum.) K. Schum.
H. pusilla Ridl.
H. scyphifera (Koenig) Steud. var. *fusiformis* Holtt.
H. scyphifera (Koenig) Steud. var. *grandis* (Ridl.) Holtt.
H. scyphifera (Koenig) Steud. var. *scyphifera*
H. striolata Ridl.

Kaempferia
K. elegans (Wall.) Bak.
K. galanga L. *
K. pulchra Ridl.
K. rotunda L. *

Plagiostachys
P. albiflora Ridl.
P. lateralis (Ridl.) Ridl.
P. mucida Holtt.

Scaphochlamys
S. atroviridis Holtt.
S. biloba (Ridl.) Holtt. var. *biloba*
S. biloba (Ridl.) Holtt. var. *lanceolata* (Ridl.) Holtt.
S. breviscapa Holtt.
S. burkillii Holtt.
S. concinna (Bak.) Holtt.
S. erecta Holtt.
S. grandis Holtt.
S. klossii (Ridl.) Holtt. var. *glomerata* Holtt.
S. klossii (Ridl.) Holtt. var. *klossii*
S. klossii (Ridl.) Holtt. var. *minor* Holtt.
S. kunstleri (Bak.) Holtt. var. *kunstleri*
S. kunstleri (Bak.) Holtt. var. *rubra* (Ridl.) Holtt.
S. lanceolata (Ridl.) Holtt.
S. longifolia (Ridl.) Holtt.
S. malaccana Bak.
S. oculata (Ridl.) Holtt.
S. pennipicta Holtt.
S. perakensis Holtt.
S. rubromaculata Holtt.
S. sub-biloba (Burk. *ex* Ridl.) Holtt.
S. sylvestris (Ridl.) Holtt.
S. tenuis Holtt.

Zingiber
Z. aurantiacum Theilade
Z. chrysostachys Ridl.
Z. citrinum Ridl.
Z. curtisii Holtt.
Z. elatior (Ridl.) Theilade
Z. fraseri Theilade
Z. gracile Jack

Z. griffithii Baker
Z. kunstleri Ridl.
Z. montanum (Koenig) Theilade
Z. multibracteatum Holtt. var. *multibracteatum*
Z. multibracteatum Holtt. var. *viride* Holtt.
Z. officinale Roscoe var. *officinale* *
Z. officinale Roscoe var. *rubrum* Theilade
Z. ottensii Val. *
Z. petiolatum Theilade
Z. puberulum Ridl. var. *chryseum* (Ridl.) Holtt.
Z. puberulum Ridl. var. *ovoideum* (Ridl.) Holtt.
Z. puberulum Ridl. var. *puberulum*
Z. spectabile Griff.
Z. sulphureum Burkill *ex* Theilade
Z. wrayi Ridl.
Z. zerumbet Smith *

Glossary

Anther: The pollen-bearing organ of a flower.

Anther appendage: Of various forms, the most common types refer to the projecting part of the tissue connecting the two pollen-bearing portions of the anther which is prolonged into either a slender beak-like structure, with infolding edges, containing the upper part of the style, as in species of *Zingiber*, or else extended into a lobed, spreading, leaf-like part called the anther crest as in species of *Amomum* and *Alpinia*.

Anther spur: This refers to the basal extensions of the pollen-bearing portion of the anther, as in some species of *Curcuma*.

Anthesis: The flowering period when a flower is open and receptive to pollination.

Apex: Tip of organ.

Aril (or arillus): An additional covering of a seed, growing out from the seed-stalk, and which may be pulpy or fleshy.

Axillary placentation: Of ovules borne centrally in a compound ovary.

Blade: Lamina.

Bract (floral bract): A much-reduced or modified leaf in whose axil a flower or partial inflorescence arises. See page 15.

Bracteole: Secondary bract situated on the pedicel below the flower. See page 15.

Bulbil: A small fleshy bud formed in the axil of a leaf or bract, which may fall and produce a new plant, as a vegetative (i.e., non-sexual) method of reproduction. The new plant will not vary but have all the features of the parent plant.

Glossary

Calyces: Plural of calyx.

Calyx: The outer circle of floral envelopes, composed of the sepals which may be free or fused together into a calyx cup or tube.

Capsule: A dry fruit that usually splits when ripe.

Central placentation: Ovules borne in the centre of the ovary.

Cincinnus: A helicoid cyme, the partial inflorescence of the Zingiberaceae. See page 15.

Coma: The complement of uppermost sterile bracts of an inflorescence that are often larger and brightly coloured, found in some gingers, e.g., *Curcuma* spp.

Coma bracts: The individual bracts forming the coma.

Connective: The sterile tissue in the centre of the anther between the pollen sacs.

Corolla: The inner circle of floral envelopes, consisting of the petals of a flower as a whole. The petals may be free or fused together to form a corolla tube with their upper free ends as lobes.

Dehiscing: Splitting or opening (of the fruit).

Distichous: In two rows.

Elliptic: Oval in outline, being of the same breadth at equal distances from the base and the apex.

Epiphytic plant (epiphyte): A plant growing on the surface of another (or else on some other elevated support) but does not derive water or nourishment from the tissues of the supporting plant.

Filament: The stalk of the stamen.

Glabrous: Without hairs; smooth, as in some leaf surfaces.

Glaucous: A blue-green or whitish bloom or substance that rubs off.

Globose: Spherical or globular.

Herbaceous flora: Non-woody plants or else those with little hard woody tissue, with no persistent stems above ground.

Indehiscent: A fruit that does not open.

Inflorescence: A flower cluster or flowering branch.

Infructescence: Cluster of fruits derived from an inflorescence.

Involucre: A whorl of bracts below the inflorescence.

Imbricate: Overlapping as in a tiled roof.

Lamina: Blade, the flat part of the leaf.

Lateral: At the side.

Leaf sheath: The lower part of a leaf enveloping the stem.

Ligule: Membranous projection from the top of the leaf sheath.

Lip: Labellum, in Zingiberaceae a petaloid staminode.

Locule: Compartment of a compound ovary or an anther.

Mottled: Marked with an irregular arrangement of spots or patches of colour.

Mucilage: A mixture of saccharides widely occurring in plants, capable of absorbing water, swelling and becoming slimy.

Mucronate: Terminated by a mucro (a short spine-like tip).

Mulch: A layer of any dried plant material such as leaves, grass, twigs, chipped wood or bark, etc., placed on the soil surface to prevent rapid evaporation of soil moisture.

Ovary: That portion of the female part of a flower in which develops the ovules (which later become seeds when fertilised).

Ovate: With the outline of an egg.

Ovoid: egg-shaped.

Parietal placentation: ovules borne on the wall.

Pedicel: Flower stalk.

Peduncle: Inflorescence stalk.

Perennating organ: A storage organ which may be a modified stem, root, leaves or bud which lies in the soil over the resting period, and mobilising the food reserves for new growth in the next growing period in association with vegetative reproduction or for self-renewal.

Persistent: Not falling off.

Petaloid: Petal-like.

Petiole: The part between the lamina and the leaf sheath in the Zingiberaceae.

Pistil: The female reproductive organ of a flower, normally comprising the ovary, style and stigma.

Pseudostem: A false stem comprising of overlapping leaf-sheaths surrounding a slender core of tender stem tissue.

Radical: Describing an inflorescence of the Zingiberaceae which is borne at the tip of a specialized leafless shoot that arises directly from a rhizome, and so appearing as a distinct structure at the base of the normal leafy shoots (compare Terminal).

Rhizome: An underground stem or root-stock, distinguished from a root by the presence of nodes, buds, and scale-like leaves; rhizomes bear roots.

Ribbed leaves: This refers to the surface of a leaf blade with prominent lateral veins.

Rosette: An arrangement of leaves apparently radiating from a centre and usually close to the surface of the ground.

Rugose: Covered with many ridges or wrinkles.

Scape: The leafless stalk of a flower cluster (or inflorescence) arising from the ground.

Sessile leaf: A leaf without a petiole or leaf stalk.

Stamen: A unit of the androecium or male reproductive organs of a flower, typically composed of a pollen-producing part (or anther) and a stalk (or filament), i.e., the pollen-bearing organ of a flower.

Staminode: A sterile stamen. In the ginger family, Zingiberaceae, the staminodes are often petal-like and showy.

Stigma: The part of a pistil or female reproductive organ of a flower which receives the pollen.

Style: The slender, elongated part of the pistil or female reproductive organ of a flower, between the ovary and the stigma.

Terminal: Describing the inflorescence of the Zingiberaceae which is borne at the tip of the leafy shoot (compare Radical).

Tomentose: Densely woolly.

References and General Reading

Baker, J.G. (1892) Scitamineae. In: Hook. f., *Flora of British India* 6: 198–264.

Barlow, H.S., Enoch, I. and Russel, R.A. (1991) H.F. Macmillan's Tropical Planting and Gardening (6th Edition). Malaysian Nature Society, Kuala Lumpur.

Burkill, I.H. (1966) *A Dictionary of the Economic Products of the Malay Peninsula*. 2nd Edition. Vols. 1 & 2. Ministry of Agriculture and Cooperatives, Kuala Lumpur, Malaysia.

Burtt, B.L. (1972) General Introduction to papers on Zingiberaceae. *Notes of the Royal Botanic Garden, Edinburgh* 31: 155–165.

Burtt, B.L. and Smith, R.S. (1972) Tentative key to the subfamilies, tribes, and genera of Zingiberaceae. *Notes of the Royal Botanic Garden, Edinburgh* 31: 171–176.

Classen, R. (1987) Morphological adaptations for bird pollination in *Nicolaia elatior* (Jack) Horan (Zingiberaceae). *Gardens' Bulletin, Singapore* 40(1): 37–42.

Cowley, J. and Theilade, I. (1995) *Zingiber sulphureum* (Zingiberaceae). *Curtis's Botanical Magazine* 12: 73–77.

Davison, G.W.H., Soepadmo, E. and Yap, S.K. (1995) The Malaysian Heritage and Scientific Expedition to Belum: Temengor Forest Reserve 1993–1994. *Malayan Nature Journal* 48: 133–146.

Gagnepain, F. (1908) Zingiberaceae. In: H. Lecomte (ed.), Flore Générale de l'Indochine 6: 25–121.

Holttum, R.E. (1950) The Zingiberaceae of the Malay Peninsula. *Gardens' Bulletin, Singapore* 13(1): 1–249.

―――― (1954) *Plant Life in Malaya*. Longman Malaysia, Singapore. (Paperback edition 1969) (Chapter 3: Bamboo, Ginger, and Orchid.)

Holttum, R.E. and Enoch, I. (1991) *Gardening in the Tropics*. Times Editions Pte. Ltd., Singapore.

Ibrahim, H. (1992) Malaysian Zingiberaceae: Ecological, morphological and economic aspects. *Bulletin of the Heliconia Society International* 6(1/2): 4–8.

────── (1995) Peninsular Malaysian gingers: their traditional uses. *Bulletin of the Heliconia Society International* 7(3): 1–4.

────── (1996) Isozyme variations in selected Zingiberaceae spp. *Proceedings, 2nd Symposium on the Family Zingiberaceae*: 142–149.

Ibrahim, H. and Zakaria, M. (1987) Essential oils from three Malaysian Zingiberaceae species. *Malaysian Journal of Science* 9: 73–76.

Ibrahim, H. and Larsen, K. (1995) A new species of *Globba* (Zingiberaceae) from Peninsular Malaysia. *Nordic Journal of Botany* 15(2): 157–159.

Kam, Y.K. (1982) The genus *Elettariopsis* (Zingiberaceae) in Malaya. *Notes of the Royal Botanic Garden, Edinburgh* 40(1): 139–152.

Khaw, S.H. (1995) Ginger Plants from the Belum Forest, Hulu Perak— A Potential Source of Natural Products. In: Ghazally Ismail, Murtedza Mohamed & Laily bin Din (eds.) *Chemical Prospecting in the Malaysian Forest*. Universiti Malaysia Sarawak/UNESCO.

Kirchoff, B.K. (1988) Floral ontogeny and evolution in the Zingiberales. In: P. Leins, S.C. Tucker and P.K. Endress (eds.) *Aspects of floral development*: 44–56.

Kress, W.J. (1990) The phynology and classification of the Zingiberales. *American Journal of Botany* 77: 698–721.

Larsen, K. (1996) A preliminary checklist of the Zingiberaceae in Thailand. *Thai Forest Bulletin* 24: 35–49.

────── (1980) Annotated key to the genera of Zingiberaceae of Thailand. *Nat. Hist. Bull. Siam Soc.* 28: 151–169.

Larsen, K., Lock, J.M., Maas, H. and Maas, P.J.M. (1998) Zingiberaceae. In: K. Kubitzki (ed.) Families and Genera of Vascular Plants. 4: 474–495. Springer Verlag.

Lim, S.N. (1972) Cytogenetics and taxonomy of the genus *Globba* L. (Zingiberaceae) in Malaya I: Taxonomy. *Notes of the Royal Botanic Garden, Edinburgh* 31(2): 243–269.

Loesener, Th. (1930) Zingiberaceae. In: A. Engler & K. Prantl (eds.) *Naturlich Pflanzenfamilien* 15a: 541–640.

Maas, P.J.M. (1979) Notes on Asiatic and Australian Costoideae (Zingiberaceae). *Blumea* 25: 543–549.

Merh, P.S., Daniel, M. and Sabins, S.D. (1986) Chemistry and taxonomy of some members of the Zingiberales. *Current Science* 55(17): 835–839.

Mustafa, A.M., Anita, H. and Ibrahim, H. (1996) Comparison of volatile compounds in 2 variants of *Elettariopsis triloba* (Gagnep.) Loesen. *Proceedings, 2nd Symposium on the Family Zingiberaceae*: 174–179.

Ridley, H.N. (1899) *Zingiber elatior*. *Journal of the Straits Branch Royal Asiatic Society*: 32: 130.

——— (1899) The Scitamineae of the Malay Peninsula. *Journal of the Straits Branch Royal Asiatic Society*: 85–184.

——— (1924) Zingiber. *The Flora of the Malay Peninsula* 4: 257–261. Reeve & Co., London.

——— (1925) Zingiberaceae. In: *Flora of the Malay Peninsula* 4: 233–285.

Roscoe, W. (1824–28) *Monandrian plants of the order Scitamineae, with 112 handcol. plates*. Liverpool.

Schumann, K. (1899) Monographie der Zingiberaceae von Malaisien und Papuasien. *Engler Bot. Jahrb.* 27: 259–350, t. 2–6.

——— (1904) Zingiberaceae. In: A. Engler (ed.) *Pflanzenreich* IV. 46. 458 pp. Berlin.

Sirirugsa, P. (1992). A revision of the genus *Boesenbergia* Kuntze (Zingiberaceae) in Thailand. *Nat. Hist. Bull. Siam Soc.* 40: 67–90.

Smith, R.M. (1985) A review of Bornean Zingiberaceae: I (Alpineae). *Notes of the Royal Botanic Garden, Edinburgh* 42(2): 261–314.

——— (1986) *Etlingera*: The inclusive name for *Achasma*, *Geanthus* and *Nicolaia* (Zingiberaceae). *Notes of the Royal Botanic Garden, Edinburgh* 43(2): 235–241.

―――― (1986) New Combinations in *Etlingera* Giseke (Zingiberaceae). *Notes of the Royal Botanic Garden, Edinburgh* 43(2): 243–254.

―――― (1986) A review of Bornean Zingiberaceae: II (Alpineae, Concluded). *Notes of the Royal Botanic Garden, Edinburgh* 43(3): 439–466.

―――― (1987) A review of Bornean Zingiberaceae: III (Hedychieae). *Notes of the Royal Botanic Garden, Edinburgh* 44(2): 203–232.

―――― (1988) A review of Bornean Zingiberaceae: IV (Globbeae). *Notes of the Royal Botanic Garden, Edinburgh* 45(1): 1–19.

―――― (1989) A review of Bornean Zingiberaceae: V (*Zingiber*). *Notes of the Royal Botanic Garden, Edinburgh* 45(3): 409–423.

―――― (1990) *Alpinia* (Zingiberaceae): A proposed new infrageneric classification. *Notes of the Royal Botanic Garden, Edinburgh* 47(1): 1–75.

Soepadmo, E. (1976) Ginger Plants. *Nature Malaysiana* 1(2): 32–39.

Stone, B.C. (1980) Additions to the Malayan flora, No. 7. A new *Geostachys* (Zingiberaceae) from Gunung Ulu Kali, Pahang, Malaysia. *Malaysian Journal of Science* 6(A): 75–81.

Theilade, I. (1998) Revision of the Genus *Zingiber* in Peninsular Malaysia. *Gardens' Bulletin, Singapore* 48(1–2): 207–236 (for 1996).

Theilade, I. and Mood, J. (1997) Two new species of *Zingiber* (Zingiberaceae) from Sabah, Borneo. *Sandakania* 9: 28–32.

―――― (1997) Five new species of *Zingiber* (Zingiberaceae) from Borneo. *Nordic Journal of Botany* 17(4): 337–347.

Tilaar, M., Sangat-Roemantyo, H. and Riswan, S. (1991) Kunyit (*Curcuma domestica*), the Queen of Jamu. In: Khozirah Shaari, Azizol Abd. Kadir and Abd. Razak Mohd. Ali (eds.) *Medicinal Plants from Tropical Rain Forests. Proceedings of the Conference.* May 13–15, 1991. Forest Research Institute Malaysia, Kepong.

Tomlinson, P.E. (1956) Studies in the systematic anatomy of the Zingiberaceae. *Botanical Journal of the Linnean Society* 55: 547–592.

――― (1962) Phylogeny of the Scitamineae: Morphological and anatomical considerations. *Evolution* 16: 192–213.

――― (1969) Classification of the Zingiberales (Scitamineae) with special reference to anatomical evidence. In: C.R. Metcalfe (ed.) *Anatomy of the Monocotyledons* 3: 295–302. Oxford.

Turner, I.M. (1995) A catalogue of the vascular plants of Malaya. *Garden's Bulletin, Singapore* 47(2): 642–655.

Valeton, T. (1918) New notes on the Zingiberaceae of Java and Malaya. *Bull. Jard. Bot. Buitenz.* 27: 1–176.

Wu, T.-L. (ed.) (1981) Zingiberaceae. In: *Flora Republicae Popularis Sinicae* 16 (2): 22–152. Science Press, Beijing.

Wu, T.-L. (1985) The origin of *Zingiber officinale. Agricultural Archeology* 2: 247–250.

Wu, T.-L., Wu, Q.-G and Chen, Z.-Y. (eds.) (1996) *Proceedings of the Second Symposium on the Family Zingiberaceae*. Zhongshan University Press, Guangzhou, China. 306 pp.

Zakaria, M. and Ibrahim, H. (1986) Phytochemical screening of some Malaysian species of Zingiberaceae. *Malaysian Journal of Science* 8: 125–128.

Index to Scientific and Vernacular Names
Page numbers in bold indicate pages with illustrations

A

'achasma' *Etlingera* species ... 110
Achasma ... 73
Alpinia ... 7, 14, 19, 21, 56, 57, **64**, 73, 85, 92, 94, 96, 105
Alpinia conchigera ... 9, **65**
Alpinia galanga ... 9, **9**, 10, 58, **66**, 95, 97
Alpinia havilandii ... 14
Alpinia javanica ... 25, **57**, 105
Alpinia latilabris... **ii**
Alpinia malaccensis ... **61**
Alpinia murdochii ... **58**
Alpinia mutica ... **19**, **58**
Alpinia nutans ... 62
Alpinia officinarum ... 9
Alpinia oxymitra ... **65**
Alpinia pahangensis ... **59**
Alpinia petiolata ... **64**
Alpinia purpurata ... 11, **60**
Alpinia rafflesiana ... **iv**
Alpinia scabra ... **63**
Alpinia vitellina ... 110
Alpinia zurumbet ... 62
Alpinieae ... 12, 18, 24, 26, 56, 98
Alpinieae group ... 13
Amomum ... 7, 14, 19, 21, 56, 67, 71, 97, 107
Amomum aculeatum ... **70**, 108
Amomum biflorum ... 25, 71
Amomum compactum ... 9
Amomum hastilabium ... 108
Amomum kepulaga ... 9, **70**
Amomum lappaceum ... **70**, 107, **107**
Amomum ochreum ... **69**
Amomum spiceum ... **67**
Amomum testaceum ... **68**, 107
Amomum uliginosum ... 25, **69**, 108, **108**
Araucaria tree ... 111
Artocarpus heterophyllus ... 113
Asplenium nidus ... 111
avocado ... 113

B

Boesenbergia ... 7, 13, 14, 18, 39, 110
Boesenbergia curtisii ... 39, **42**
Boesenbergia pandurata ... 40
Boesenbergia plicata ... **40**, **41**
Boesenbergia rotunda ... 9, 10, 39, 40, **42**
bonglai ... 10
bonglai hitam ... 10
buah pelaga ... 8
Burbidgea ... 18

C

Camptandra ... 14, 39, 43, 98
Camptandra latifolia ... 43
Camptandra ovata ... 43, **43**
Camptandra parvula ... 12, 25, 43
Canna ... 21
Cannaceae ... 3
cardamom ... 8, 9, 67, 71
cardamom of commerce ... 18
Catimbium ... 57
Caulokaempferia ... 20
cekur ... 9, 12, **52**, 110
Cenolophon ... 57
Cinnamomum iners ... 113
cinnamon ... 113

Costaceae ... 3, 4, 94
Costeae ... 3
Costoideae ... 3
Costus ... 3, 4, 21, 94
Costus globosus ... **5**
Costus speciosus ... **6**
Curcuma ... 7, 12–14, 18, 21, 39, 44, 53, 93, 94, 97, 103, 104, 112, 113
Curcuma aeruginosa ... 10, **45**, 113
Curcuma alismatifolia ... 11
Curcuma aurantiaca ... **44**, 113
Curcuma domestica ... 8, 10, **44**, 112, 113
Curcuma longa ... 8
Curcuma mangga ... 9, 10, 112, 113
Curcuma roscoeana ... 11
Curcuma xanthorhiza ... 10, 94, 103, 112, 113
Curcuma zedoaria ... 10, 94, 112, 113

D

Dimerocostus ... 4

E

Elettaria ... 7, 14, 56, 71
Elettaria cardamomum ... 8, 18, 71
Elettaria longituba ... 71
Elettariopsis ... 7, 13, 14, 56, 71
Elettariopsis burttiana ... 1
Elettariopsis curtisii ... 9, **71**, 100, 109
Elettariopsis smithiae ... 1, 100, 109
Elettariopsis smithiae var. *rugosa* ... 100, 109
Elettariopsis triloba ... 56, **72**, 95, 109
Etlingera ... 7, 14, 18, 20, 25, 56, 73, **78**, 105

Etlingera elatior ... **x**, 9, 11, 12, 19, **20**, 75, **82**, **83**, **84**, 106
Etlingera hemisphaerica ... **83**, 106
Etlingera littoralis ... **75**, **76**, **105**, 106
Etlingera maingayi ... **80**, **81**, 106, 107
Etlingera metriocheilos ... **73**, 106
Etlingera 'nicolaia' group ... 106
Etlingera pauciflora ... 106
Etlingera punicea ... **2**, 25, **77**, 106
Etlingera triorgyalis ... **74**, 99, 106
Etlingera venusta ... **78**, **79**, 106

G

Geanthus ... 73
Geocharis ... 14, 56, 85
Geostachys ... 14, 25, 56, 85, 98
Geostachys decurvata ... **87**
Geostachys elegans ... **86**
Geostachys leucantha ... 1
Geostachys cf. *megaphylla* ... **85**
Geostachys taipingensis ... **86**
Globba ... 10, 12, 18, 26, 94, 96, 97, 98, 101
Globba cernua ... 100, 101
Globba cernua complex ... 26
Globba corneri ... **27**
Globba fragilis ... 1, **29**
Globba holttumii ... 1
Globba leucantha ... **30**
Globba nawawii ... 1, **28**
Globba patens ... **30**
Globba pendula ... 25, **29**
Globba schomburgkii ... **20**, **30**
Globba unifolia ... **28**
Globba variabilis spp. *pusilla* ... **27**
Globba winitii ... 11
Globbeae ... 18, 24, 26, 98
greater galangal ... 9

H

halia ... 8, 32
halia bara ... 32, 93
halia padi ... 32
Haniffia ... 24, 39, 46
Haniffia cyanescens ... 46
Haplochrema ... 18
Hedychieae ... 12, 13, 18, 24, 26, 39, 98
Hedychieae group ... 13
Hedychium ... 7, 20, 25, 39, 46, 47, 94, 103
Hedychium collinum ... **46**
Hedychium coronarium ... 11, 47, **50**, 113
Hedychium longicornutum ... 25, **47**, **48**, 100, 103, **104**, 111
Hedychium paludosum ... **49**
Heliconiaceae ... 3, 19
Hornstedtia ... 12, 56, 87, 97, 98
Hornstedtia conica ... 109
Hornstedtia ophiuchus ... 100, 108, 109
Hornstedtia scyphifera ... **88**
Hornstedtia scyphifera var. *grandis* ... **89**, **90**
hornstedtias ... 112

J

Jamu ... 10

K

Kaempferia ... 7, 12, 13, 18, 21, 39, 51, 97, 98, 113
Kaempferia galanga ... 9, 10, 12, 51, **52**, 110
Kaempferia pulchra ... 11, **11**, **52**, 110
Kaempferia rotunda ... 51, **51**

kantan ... 9, 12, 106
kayu manis ... 113
kunyit ... 112
kurkum ... 44

L

laksa ... 9
laksa asam ... 9
Languas ... 57
lempoyang ... 10
lempoyang hitam ... 10
lengkuas ... 9
Lowiaceae ... 3

M

Maranta ... 21
Marantaceae ... 3, 13
minor galangal ... 9
Monocostus ... 4
Musaceae ... 3

N

nangka ... 113
nasi kerabu ... 9
nasi ulam ... 9
Nicolaia ... 73
'nicolaia' *Etlingera* species ... 110

O

orchids ... 18

P

Persea americana ... 113
perut ikan ... 9
Plagiostachys ... 14, 56, 92, 97
Plagiostachys albiflora ... **92**
Porcelain Flower ... 75
Pteridophytes ... 23

R

rambutan ... 107
rendang ... 9
Riedelia ... 85
Roscoea ... 11, 46
Rubiaceae ... 22

S

Scaphochlamys ... 13, 25, 39, 53, 104, 110
Scaphochlamys concinna ... **55**
Scaphochlamys longifolia ... **55**
Scaphochlamys kunstleri ... **54**, 100, 104, **104**
Scaphochlamys sub-triloba ... **53**
Scitamineæ ... 22
Siamese Tulip ... 11
singabera ... 1
Strelitziaceae ... 3

T

Tapeinochilos ... 4
temu hitam ... 10
temu kuning ... 10, 112
temu lawak ... 10, 103, 112
temu pauh ... 10, 112
Torch Ginger ... 11, 75
turmeric ... 9, 44

U

ulam ... 7, 10

V

Valanga nigricornis ... 114

Z

zanjabil ... 1
zedoary ... 10

Zingiber ... 1, 7, 12, 13, 14, 25, 26, 31, 32, 93, 94, 97, 98, 101, 102, 103, 113
Zingiber citrinum ... **32**
Zingiber curtisii ... 102
Zingiber fraseri ... 1, **34**
Zingiber gracile ... 102, 110
Zingiber gracile complex ... 32
Zingiber griffithii ... **34**, 110
Zingiber kunstleri ... **34**, 110
Zingiber montanum ... 10
Zingiber multibracteatum ... **32**
Zingiber officinale ... 1, 8, 10, 32, 93, 94, 95, 112
Zingiber ottensii ... 10, **33**, 93, 110
Zingiber puberulum ... 14, **33**, **34**, 102
Zingiber puberulum group ... 32
Zingiber spectabile ... **viii**, 11, **35**, **36**, 94, 95, 101, **102**
Zingiber suphureum ... 1, **38**
Zingiber zerumbet ... 9, 10, **31**, 93, 103
Zingiberaceae ... 1, 3, 4, 7, 8, 10, 14, 18, 19, 21, 22, 23, 24, 93, 94, 96, 97, 99, 110
Zingiberales ... 3
Zingibereae ... 18, 24, 26, 31, 98
zingiberi ... 1
Zingiber officinale ... **7**, 32, 95

Other titles by *Natural History Publications (Borneo)*

For more information, please contact us at

Natural History Publications (Borneo) Sdn. Bhd.
A913, 9th Floor, Wisma Merdeka
P.O. Box 13908, 88846 Kota Kinabalu, Sabah, Malaysia
Tel: 6088-233098 Fax: 6088-240758 e-mail: chewlun@tm.net.my

Mount Kinabalu: Borneo's Magic Moutain—an introduction to the natural history of one of the world's great natural monuments *by* K.M. Wong & C.L. Chan

Enchanted Gardens of Kinabalu: A Borneo Diary *by* Susan M. Phillipps

A Colour Guide to Kinabalu Park *by* Susan K. Jacobson

Kinabalu: The Haunted Mountain of Borneo *by* C.M. Enriquez (Reprint)

A Walk through the Lowland Rainforest of Sabah *by* Elaine J.F. Campbell

In Brunei Forests: An Introduction to the Plant Life of Brunei Darussalam (Revised edition) *by* K.M. Wong

The Larger Fungi of Borneo by David N. Pegler

Pitcher-plants of Borneo *by* Anthea Phillipps & Anthony Lamb

Nepenthes of Borneo *by* Charles Clarke

The Plants of Mount Kinabalu 3: Gymnosperms and Non-orchid Monocotyledons by John H. Beaman & Reed S. Beaman

Slipper Orchids of Borneo *by* Phillip Cribb

The Genus Paphiopedilum (Second Edition) *by* Phillip Cribb

Mosses and Liverworts of Mount Kinabalu *by* Jan P. Frahm, Wolfgang Frey, Harald Kürschner & Mario Manzel

Birds of Mount Kinabalu, Borneo *by* Geoffrey W.H. Davison

Proboscis Monkeys of Borneo *by* Elizabeth L. Bennett & Francis Gombek

The Natural History of Orang-utan *by* Elizabeth L. Bennett

The Systematics and Zoogeography of the Amphibia of Borneo *by* Robert F. Inger (Reprint)

A Field Guide to the Frogs of Borneo *by* Robert F. Inger & Robert B. Stuebing

The Natural History of Amphibians and Reptiles in Sabah *by* Robert F. Inger & Tan Fui Lian

A Field Guide to the Snakes of Borneo *by* Robert B. Stuebing & Robert F. Inger

Marine Food Fishes and Fisheries of Sabah *by* Chin Phui Kong

Land Below the Wind *by* Agnes N. Keith (Reprint)

Three Came Home *by* Agnes N. Keith (Reprint)

An Introduction to the Traditional Costumes of Sabah (*eds.* Rita Lasimbang & Stella Moo-Tan)

Manual latihan pemuliharaan dan penyelidikan hidupan liar di lapangan *by* Alan Rabinowitz (*Translated by* Maryati Mohamed)

Etnobotani *by* Gary Martin (*Translated by* Maryati Mohamed)